130 YEARS OF CATCHING UP WITH THE WEST

To my late grandfather, Ludwig Maurer,
for teaching me about politics

The Author would like to thank the Austrian Federal Ministry of
Science and Transport for their generous support.

130 Years of Catching Up With the West

A comparative perspective on Hungarian industry, science and technology policy-making since industrialization

PETER S. BIEGELBAUER
Institute for Advanced Studies, Vienna

Routledge
Taylor & Francis Group

LONDON AND NEW YORK

First published 2000 by Ashgate Publishing

Reissued 2021 by Routledge
2 Park Square, Milton Park, Abingdon, Oxon, OX14 4RN
52 Vanderbilt Avenue, New York, NY 10017

Routledge is an imprint of the Taylor & Francis Group, an informa business

Publisher's Note
The publisher has gone to great lengths to ensure the quality of this reprint but points out that some imperfections in the original copies may be apparent.

Disclaimer
The publisher has made every effort to trace copyright holders and welcomes correspondence from those they have been unable to contact.

A Library of Congress record exists under LC control number:

ISBN 13: 978-1-138-63452-7 (hbk)
ISBN 13: 978-1-138-63447-3 (pbk)
ISBN 13: 978-1-315-20532-8 (ebk)

Contents

List of Figures

List of Tables

Preface

By the time my grandfather reached my age of 33, he had lived through four distinctly different forms of societal organization. The first of these forms was the Austro-Hungarian Monarchy, the second the unlucky, democratic First Republic of Austria and the third the undemocratic Austro-fascist 'Ständestaat'. By 1940, he was living in the dreadful era of totalitarian national socialism, fighting a war he did not believe in, for a 'Volk' and a 'Führer' he at the same time feared and hated. In the midst of the 'Blitzkrieg', he neither could have known that the war was only in its early stages, nor that he would survive the war and have the opportunity to start a new life in the democratic neo-corporatist Second Republic of Austria.

My grandfather told me many stories about the old days. He never found anything positive about the old times, other than the fact that he and my grandmother had been young. In fact, he always had hoped that people would learn from history, although over time his doubts seemed to increase. It must have been because of his stories that I discovered an interest in how societies organize themselves – or are organized by others.

The topic of this study is how a country organized parts of its society during a time span of 130 years, namely the way the Hungarians structured the science and technology system of their society since industrialization. I decided to analyze this system, because I think it is of crucial importance to any society in respect to its culture and its social, political and economic structures. I decided to analyze the Hungarian science and technology system, because of the many changes in societal organization this country went through in the course of these 130 years – deeper and more fundamental changes even than the ones Austria experienced.

In fact, during the last 130 years Hungary experienced a wide variety of regimes, among them absolute-monarchical, enlightened-monarchical, fascist, pluralist Western-democratic, realsocialist, realsocialist-reformist and Soviet style governments. During these turbulent years, the Hungarian science and technology system changed as well. These interlinked changes, which took place in one, constantly shrinking, geographic area, make Hungary an interesting case for a time-series study. In this type of study one has the possibility to study changes of countries or more broadly systems, not across countries, but across time-periods.

The effort seems justified as anyone who wants to reform a national economy, has to bear in mind that the country in question has been formed by a long line of historical events, still influencing the present, even if the original stimulus has vanished already. Accordingly, Hungarian society still bears the imprint of the political systems it featured during the last decades, if not centuries. Therefore, any analysis of the problems facing Hungary today has to begin with a discussion of the country's history. This form of analysis allows one to view the current reform movements in light of the pitfalls of similar, perhaps not so far reaching reform movements of the past.

This book is a historical analysis of the development of the Hungarian S&T system with a comparative perspective on Austria, the Czech Republic, Germany, Poland, Slovakia and Slovenia. This study has a strong interdisciplinary approach and is based on the secondary analysis of existing quantitative data. The data was derived from research done primarily at the libraries of the Massachusetts Institute of Technology (MIT) and Harvard University in Cambridge, Massachusetts, in the United States; the library of the Hungarian Central Statistical Office and the National Technical Information Center and Library (OMIKK) in Budapest, Hungary; as well as the libraries of the Austrian East and Southeast Europe Institute (ÖOSI), the library of the University of Vienna and of the Institute for Advanced Studies in Vienna, Austria.

The information described above is complemented by approximately 120 expert interviews conducted through structured interviews of a number of researchers (economists, political scientists, sociologists, engineers) in Austria, Hungary and the United States over the course of five years. A large number of prepared informal interviews was done primarily with individuals from the S&T sphere in Budapest and Vienna.

Moreover, information was drawn from the Internet using various methods. The Hungary discussion list of George Washington University, Washington, D.C., and the Central and Eastern Europe list of the University of Economics in Vienna provided addresses, contacts and at times discussion fora. A number of sites on the World Wide Web were utilized for gathering information. The electronic weekly 'Hungary Report', which was published by Hungary-Online until 1997, was particularly helpful as a digest of the leading Hungarian dailies.

I could not have written this thesis alone. I had the help of many people in four different countries. First of all, I am indebted to the late Charlotte Teuber and Otmar Höll from the Political Science Institute at the University of Vienna, who were my dissertation readers and helped me overcome all kinds of barriers. Charlotte was also a rich source of knowledge about social science and life as such, who over time became a good friend of mine. Andreas Pribersky from the Austrian East and Southeast European Institute (ÖOSI) provided me with crucial help and important resources both in Vienna and Budapest. Janos Fath made valuable comments on a number of exposés of this book. I also want to thank Josef Leidenfrost and Florian Gerhardus from the Austrian Academic Exchange Service (ÖAD) for information on the subject. I want to thank Ronald Pohoryles and Liana Giorgi from the Interdisciplinary Center for Comparative Research in the Social Sciences (ICCR) for supervising my work in Slovenia and their patience with me during the time I wrote my dissertation, which, together with my master's thesis, written in the United States, formed the basis for this book.

In the United States I want to thank Eugene Skolnikoff, my academic and thesis advisor at the Political Science Department at the Massachusetts Institute of Technology (MIT). From him I received critical and often very candid comments on my academic efforts. I also want to thank Nick Ziegler from the Sloan School of Management, who was the second reader of the thesis, and who provided valuable materials during a related course at MIT. Henry Ergas, during his stay at the Kennedy School of Government at Harvard University, has guided me through much of the first half of my research. I also want to thank Richard Locke from MIT and Mária Kovács from Harvard University for comments on earlier exposés.

In Hungary, two of the most experienced researchers of Hungarian science and technology, Anamária Inzelt at the Innovation Research Center (IKU) and Katalin Balázs at the Institute of Economics of the Hungarian Academy of Sciences (MTA), were of great help when they shared some of their research results with me. The Institute for Research Organization of the MTA turned out to be an information gold mine for my work: Judith Mosoni, György Darvas, András Füzesséry, and Mariann Tarnóczy provided ample information. Furthermore I am indebted to Peter Szántó from the National Technical Information Center and Library (OMIKK), who provided help when I was in despair because all my prospective interview partners were on holiday during the summer of 1994. Istvan

Szemenyei, from the National Committee for Technical Development (OMFB), went through an earlier version of this thesis and interviews in Budapest and Washington, D.C. Imre József and Árpád Jelinkó, both from OMFB; Éva Sztankó from the Ministry of Industry, Trade and Tourism; Tamás Kemény from the Hungarian Industrial Foundation; Frigyes Geleji from the Bay Zoltán Foundation; Pál Tamás from the Institute for Sociology of the MTA as well as Tibor Braun of the Library of the MTA also had to endure a number of interviews. Last, but not least, Gábor Palló from the MTA's Institute of Philosophy was a critical voice, providing his viewpoints during a number of personal conversations.

Gábor was also part of a group of Hungarian and German academics writing about the history of scientific and technological relations between Germany and Hungary. I was provided with the opportunity to cooperate with this group from 1997 on and I received important commentaries from a number of members, especially from the coordinator of the group, Holger Fischer, from the Institute for Finno-Ugrist Studies at the University of Hamburg.

Generous funding for my studies in the United States and Hungary were provided by the Austrian Economic Chamber (WKÖ) and additional support came from the Siegfried Ludwig Fund, the Austrian Federal Ministry for Science and Research (BMWV), the University of Vienna, the Julius Raab Foundation and the Windhag Foundation. Funding for my preliminary studies of the Hungarian science and technology policies, which finally lead to this thesis, was provided by the ÖAD. In addition, the BMWV provided funds for the translation of this book into English.

I am also indebted to Gudrun Anreither-Stenzenberger, Sara Caporal and Gertrud Hafner, who edited the book with regards to language and layout.

Finally, and most importantly, I want to express my thanks to my late grandfather, Ludwig Maurer, who, in one way or another, was always with me during the last years.

Peter Biegelbauer
Institute for Advanced Studies,
Vienna, Austria

1 Introduction

One way to understand and to learn from the past is by viewing history as a series of innovations, which were introduced to all societies with varying levels of success. Indeed, one might measure the pace of a society's evolution by analyzing the number and nature of successfully introduced innovations. And in fact, such an understanding of history is often implicitly employed by historians of science and technology even if they are not always explicit about their normative basis of assumptions.

One example is 15th century Ming dynasty China, a society which seriously restrained its technological capabilities in an important area, when it, abandoned deep sea shipbuilding technologies for internal reasons (Pacey 1990). Of course, at the same time China made important contributions in other fields, as for example in that of philosophical thought. However, it seems that these advances did not strengthen the country's scientific, technological and economic capacities, which indeed stagnated or declined for a long period of time. This economic and technological stand-still made isolationist China more vulnerable to other peoples, specifically the Europeans, who at this time began to systematically explore the world and whose subsequent colonial activities proved to be as disastrous for Asia as for other parts of the world.

Contrary to this, an example for the successful introduction of innovations to a society, is 18th century England, in which individual entrepreneurship, competition and the development of what could be called a 'civil society' of people less dependent on the church and on the state, enhanced the possibility of introducing innovations at an up to then unprecedented speed (Rosenberg/Birdzell 1986). Of course, at the same time a social phenomenon occurred, which was unprecedented in its dexterity, too: the pauperization of a large part of society and the creation of the industrial proletariat (Polanyi 1944). Yet, as severe as these social developments were, they did not prevent England from becoming the first global hegemon for most of the 19th and the first part of the 20th century.

Following from this, 'history' should not be understood as a solid concept, but rather as consisting of a plurality of 'histories', which are defined by the specific nature of the unit of analysis. If the history of

1

technology of China or England is analyzed, it will allow for a very different perspective of the past, than the social or economic history of the respective countries. What makes the situation complicated is that these different histories are interlinked with each other to varying degrees, depending on the circumstances of specific time periods. Without the successful introduction of the steam engine in England, the industrial revolution might not have been all that revolutionary, but might have taken a much slower and evolutionary path. In this case of the steam engine a technological innovation had a strong impact on the economy of a country, which entailed a multitude of often quite indirect effects ranging from the rising political power of the country to the declining social situation of the factory workers. Therefore, an incident, which might be of importance for 'technological history' might have effects on 'economic history', 'political history' and 'social history'.

Such observations have led to the hypothesis that technological change might drive history (Heilbronner 1967). Others have rightly pointed out that technological innovations themselves might be contingent on social factors, or indeed, social innovations. This in turn has led to the hypothesis that society in fact influences, or even shapes, technological change[1] (Williams 1994). Coming back to the example of the steam engine, Rosenberg and Birdzell (1986) identify a number of social factors responsible for the success of Europe in comparison to other regions of the world, amongst which we find the diminishing of the church's influence on political decisions and the fact that the European states were in competition to each other politically, economically and technologically. Following this logic, the successful introduction of the steam engine might have been possible only because a number of social innovations had occurred before, such as the strengthening of civil society and the individual entrepreneur. The role of the entrepreneur was of central importance here, since he was forced to take risks and invest into new technologies to enlarge his profit margin because of the rising level of competition in late 18th century England.

From these statements it can be inferred that it is actually quite difficult to establish clear causal relationships between various 'histories' and between various innovations. One might even surmise that the question as to whether technological change drives history or whether society shapes technologies is irrelevant, because of the interlinked nature of both processes. This suspicion is nourished by the observation that not only the concept of 'history' needs a number of qualifications, but that the notion of 'innovations' similarly incorporates a number of facets, which should be differentiated.

If we are interested in learning about history and from history and we accept that history can be viewed as a series of innovations, we have to specify the nature of innovations in more detail. In fact, it is sometimes difficult to specify innovations, as they appear in different forms, sizes and kinds. These innovations might take the forms of new products, such as the wheel, the automobile or the computer, new processes, such as the introduction of bronze, the hot-bed molding of iron or the numerically controlled workbench and new organizational set-ups, such as the creation of the nation state, the manufacture or corporate alliances. The innovations might come in different sizes, as major, radical innovations, such as the radio and the Internet, and minor, incremental innovations, such as the adaptation of a 220 Volt powered German machine tool to the United States' 120 Volt electrical power sources or the production of tyres with longer life cycles due to a different mixture of largely the same chemical components already used before. The innovations may also come in various kinds, being generated in a number of areas as different as the political sphere, where new voting systems might be recognized as more democratic, or the social sphere, where new forms of formerly societally unrecognized work may reduce unemployment rates, or the economic sphere, where environmentally sound production processes may lead to sustainable forms of development, to name only a few examples.

Having understood the spectrum of existing forms of innovation, it soon becomes clear that the causal relationship is not fixed; neither between the different kinds of innovations, in the sense of technological, social or political, nor between different forms of innovations, in the sense of product or process, radical or incremental. On the one hand, the industrial revolution was not only an effect of a number of product and process innovations such as the steam engine and new weaving and dyeing techniques; it was also preceded by a societal reorganization in the 18th century freeing a part of the labor force from slavery in the agricultural sector (Polanyi 1944). On the other hand, the ongoing computerization of the OECD countries starting in the 1970s, which has been changing the organizational structure of all highly industrialized societies decisively, was preceded by the invention, development and dissemination of a bundle of microelectronics technologies (Bell 1973).

If the current effort is to learn about history and from history, even in perhaps quite a limited way, a serious analysis of the interrelationship between the different forms of innovations cannot be simplistic. In fact, it shall be argued here that the respective contingencies of historical development are as complicated as they are badly understood. Taking into account the level of complexity of the subject, it seems hardly feasible to create an empirically based analysis of the intertwined relationship of the political, social, cultural, economic, scientific and technological innovations in histories of industrialized societies by addressing all these different areas at once. Therefore, this study shall focus on the interaction of political, economic, scientific and technological innovations in the area of Central Europe from the time of this region's industrialization in the 1860s until the end of the 20th century. More specifically, this book is about the interdependency between the political sphere and the areas of science and technology (S&T) as well as, on a more general level, the economy in Hungary in comparison to other Central European countries particularly Austria and Germany, from the 1860s to the 1990s.

On another, more reflexive, level of analysis not only will the understanding of the complex relationship between the different kinds and forms of innovations, but also the coming to an insight as to how people in different time periods understood the afore mentioned complex relationship shall be at the focus of interest. Limiting the task to a more concrete assignment, the gradual development of people's ideas about technological change and their ideas on how to govern the processes of change will be analyzed, taking the Hungarian experiences as the example for a comparative perspective from the 1860s to the 1990s.

The analysis of people's notions of technological change and their perspectives on questions of governance of this process is highly significant insofar as it should give us deeper insights into the way people perceive history (or single 'histories' in the sense of technological or economic history) and into the possibilities of influencing historical processes in order to understand our own inefficiencies in thinking about these questions. A better comprehension of these problems is indeed important for the theoretical understanding of the linkages between societal development and technological change, but also simply for the practical reason of actual policy-development in the areas of economic, science and technology policies.

The systematic study of the impact of ideas and notions about the functioning of the world in general (or about more specific problems such as the processes of technological change) on the development and delivery of policies has been rather neglected up to now. More generally, it can be stated that the interest in the effect which ideas have on the behavior of decision-makers is commonplace in some social science disciplines – the work of Thomas Kuhn and others (Kuhn 1970, Cohen 1985) on the sociology of science comes into mind – while this appears to be less the case for others, as for example political science and economics.

In fact, in these two latter disciplines ideas' effects on the behavior of actors have so far only been studied rather implicitly. One example from political science is the concept of political culture (Almond/Verba 1963), which analyzes the impact of these norms, values and daily routines, which are generally perceived as being part of a specific 'culture', on the set-up of political entities, called polities, and on politics as such. Yet, although norms and values are amongst the basic ingredients for ideas and notions about the functioning of a particular facet of life, the two concepts are not equivalent. In fact, one of the most visible differences between basic norms and values and more complex ideas and notions is that the former tend to resist change and evolve very slowly, whereas the latter have life-cycles of sometimes only two decades.

Another example is the economic approach to human behavior (Becker 1976), which operates on the basis of the assumption that all human beings act rationally, that their preferences do not change substantially over time and that markets, which are defined very broadly here (the economic stock market as well as the social marriage 'market'), tend to move towards equilibria. With these assumptions in mind, the approach aims at an explanation of human behavior as 'the allocation of scarce means to satisfy competing ends' (Becker 1976, p.3). While norms and values can enter the model via the concept of preferences, more complex categories such as ideas and notions are of no explicit interest for social scientists using this approach.

Of course Becker's work has a number of forerunners. In a long-ongoing effort political philosophers like for example 16th century Niccolò Macchiavelli in his 'Il Principe', have interpreted single policies and, more broadly, politics as such as a process with the ultimate goal of obtaining power. Political scientists and economists in the tradition of the utilitarianism of Jeremy Bentham, the historical materialism of Karl Marx or the historicist statism of Friedrich List have learned to analyze policies and politics as a process with the goal of maximizing an individual's, a class's or a nation's utility, be it in the form of money or power, but also truth or other categories.

Besides these analyses of the question as to for what reason and to which end individuals or groups want to enter politics or economics, the question as to how they actually engage into politics and the economy is relatively well understood, too. These aspects are covered by research into the processes of policy-development and policy-delivery as well as other forms of governance, such as the development of management strategies and the actual delivery of the same. Examples for successful research along these lines are quite diverse, as may be demonstrated taking the example of the analysis of political interest formation, which has been described by liberal group theory (Truman 1971), by literature on neocorporatism (Schmitter 1979), by the new political economists (Downs 1957) and by others.

A topic which has not been given much thought until now is the way decision-makers' inherent ideas or concepts influence the policies they devise. A study of this topic includes the analysis of the cognitive background of the decision-finding and -making processes, which was described and analyzed by economists and political scientists from a number of traditions, including the ones cited in the previous paragraph. Existing research, however, tends to exclude the very ideas and notions which form the framework through which the decision-makers interpret the problems they face and subsequently structure their problem-solution strategies.

This said, it is important to mention a small strand of literature in political economy and international relations, which has been growing throughout the 1990s. In both subfields several books have been published, which are part of neo-institutionalism in both its historic (Hall 1993) and rationalist (Goldstein/Keohane 1993) variants (compare with Jacobsen 1995 and Blyth 1997). Moreover, in Germany Hofmann (1993) and Braun (1997) have analyzed the roles of ideas with regards to the steering problems modern states face in societies of growing complexity.

Most of these studies use ideas only as additional or intervening variables for explanations of policy development (Blyth 1997). The work of Peter Hall (1986, 1989, 1993) forms an exception; for Hall, the importance of the 'ideational' (sic!) factor seems to grow over the years: in his earlier work, he deems ideas to attain importance primarily through their inclusion in institutions (1986, pp.278), later, ideas are understood as of considerable importance on their own right (1993, p.289).

For the sake of clarifying what is at stake, one might say in somewhat simplifying terms that political science and economics have obtained quite a clear vision of the 'know-why' and of the 'know-how', but much less of the 'know-what' of the political and economic processes. In other words, the reasoning behind the entrance of individuals and groups into politics and the economy as well as the governance of the two areas are well-understood, but the cognitive basis upon which individuals and groups in politics and the economy form their decisions, leading to policies, management strategies and the like are barely understood.

It seems to be the case that decision-makers, at least in specific policy-fields and time-periods, have a widely-shared understanding of the general mechanisms behind the problems they face. This general understanding, which is based on the repeatedly mentioned ideas and notions underlying policies, can take the form of paradigmatic notions and shall therefore be called 'policy paradigms' here.[2] These policy paradigms are not equivalent to, and may even be independent of, ideologies or 'Weltanschauungen'.

Policy paradigms are cognitive frameworks common to a number – indeed, most of the time to the overwhelming majority – of policy-makers in a specific time-period and policy-field. They consist of the general understanding of the key process(es) in the center of interest of a specific policy-field, such as the process of technological change in S&T policy or the main mechanisms behind economic cycles of boom and bust in macro-economic policy-making. Moreover, they consist of the broadly defined goals of policies in this area and entail an equally broad understanding of the set of policy-instruments decision-makers have at their hands to tackle the problems of a field of public policy. In a sense the policy paradigms even enable policy-makers to develop policies as they create an intersubjective view of specific problem areas between groups of decision-makers and therefore form a basis for discussions.

The term 'policy paradigm' has been used independently by both Peter Hall (1993) and Beatriz Ruivo (1994), for whom they carry different meanings (for a more detailed statement, see the annex). Hall's concept, which is closer to the usage of the term here, has focused on the impact of economic theories on economic policy-making, whereby France and Great Britain served as case studies, since in both countries' governments

Keynesian macroeconomic policies were exchanged against Monetarist ones in the 1970s and 1980s. Because of the different policy fields, which are the focus of Hall's study and of this book here, we find a few differences in the way in which the term 'policy paradigm' is applied in the two studies. The same holds true for a Beatriz Ruivo's usage of the term as compared to this study. Ruivo applies the term 'policy paradigm' to science policies, which comes close to this book's subject of interest, yet defines a set of policies – and not the notions and cognitive models underlying these policies – as a paradigm.

Existence and nature of these policy paradigms shall be analyzed in the course of this study on the example of Hungarian S&T policy-making since the time of industrialization. Since Hungary and indeed all Central Europe experienced a number of different forms of political and economic regimes in the 130 years which will be analyzed here, the country ought to be a good example for the study of the nature of policy paradigms.

The respective questions to be studied in this book are: What are the underlying notions for a specific set of policies? How have they become important enough to be considered for policy development and where did they come from? Only with the help of a comprehensive understanding of these questions will it be possible to really grasp the mechanisms of policy-development and the underlying reasons for policy successes and failures.

Another theme, which shall be important throughout the book: the evaluation of policies in the sphere of science and technology (S&T), but also in fields, which are linked to these areas, most importantly economic and educational policies, which were employed in Hungary during the analyzed time period. This effort is important for two reasons: On the one hand, it will provide a possibility to learn more about the ways in which the ideas and notions underlying specific policies are subject to change and also about the reasons for change. On the other hand, by identifying the reasons for policy failures, an attempt will be made to identify lessons for decision-making in the respective policy fields.

Therefore the three major themes of the study shall be the following: a process oriented understanding of history incorporating various perspectives, in which the analysis of the intertwined political, economic, scientific and technological 'histories' of Hungary, Austria, Germany and a few Central European countries shall be at the focus of interest; the analysis of the changing comprehension of the interlinkage of political, economic, scientific and technological factors and changes in the ideas and paradigmatic notions underlying the policies directed at the S&T sphere; finally, as an evaluatory aspect linked closely to the two other questions, the reasons for success and failure of the steering efforts over the period of time in question shall be analyzed.

As has been pointed out, these questions shall be analyzed in the form of a time-series study of Hungary, with frequent comparisons to other Central European countries, Austria and Germany. The comparative aspect of the study is all the more important since historical analyses of S&T in the region have shown that the countries in discussion, due to their small size and long common borders, are heavily dependent on each other (Fischer/Szabadvary 1995, Biegelbauer 1997).

Hungary, like the other countries in Central and East Europe, has been economically less well developed than the Western part of the continent for hundreds of years. With the industrialization of the 19th century, the disparities between the Western and Eastern part of the continent grew (Rokkan 1987, Kennedy 1987), sometimes ignoring the borders of the countries. An especially striking case is the Austrian Habsburg Empire, where the Czech and German speaking Northwestern and Western areas were quite well developed, whereas the Eastern and Southeastern parts of the empire, including Hungary, were comparatively backward (Gerschenkron 1977). With being granted the partial sovereignty of Hungary and the construction of the Dual Austro-Hungarian Monarchy in 1867, the Hungarian leaders, politicians, scientists and economists alike, were confronted with a viable chance to develop their national economy rapidly for the first time in Hungarian history. And indeed, as will be shown in the course of the book, the central motif for many Hungarian leaders of the time after industrialization was to economically, technologically and scientifically catch up with Western Europe.

The development of Hungarian science and technology can be divided into a number of periods. These time spans, which shall be described later, are interlinked with the historical changes which the region went through. This insight provides the possibility to understand the recent changes in Central and Eastern Europe not as isolated events, but to see them in perspective. Thus, it becomes clear that Central and East Europe as a region has been in a race against time, trying to catch up with the Western part of the continent at least since the beginning of the 19th century.

For the last 130 years a number of ideologies, of which were all introduced to Hungary in times of societal change, have been used to reach this goal of obtaining a higher level of economic development. Over time, the most influential ideologies in Hungary and in the region in general were nationalism, capitalism, fascism and realsocialism. Of course, these ideologies never have been realized in any 'pure' form, but have mostly been mixtures. For example, fascism was paired with nationalism – and to a certain degree also with capitalism, although the war economy had some features of a command economy and was therefore certainly not a market economy proper. Another example is realsocialism, which, from the early 1960s on, featured more and more elements of capitalism, and whose Hungarian variant was sometimes, a bit disrespectfully, called 'goulash communism'. Nevertheless, in each time period a single ideology structured the way in which the other ideologies were utilized and can therefore be seen as dominant.

As will be shown during the book, nationalism, capitalism, fascism and realsocialism all were utilized by ruling groups for catch-up efforts, be it for the good of the nation, the capitalists, the race, or the working class. Until the present day none of these catch-up efforts has been fully successful, although some attempts have been more effective than others. Each time a new ideology was utilized in Central and Eastern Europe (CEE), a paralleling transition from one system of societal organization to another was initialized. Therefore, when addressing the recent changes in the region following the fall of the Berlin Wall in 1989, it is wrong to speak of 'the' transition, precisely because the latest transition from realsocialism to capitalism in CEE is only one in a series of transitions in this part of the world.

Once this has been understood, it should be possible to learn from an analysis of the past transitions to better understand the latest one. Since such an effort can be successful only if the analysis of past developments is carried from the abstract and general to the exemplary and concrete level, this study analyzes the transitions of the Hungarian science and technology (S&T) system, which it has undergone since the beginning of industrialization. Here, a 'system' shall be a societal sphere, which consists of a number of persons and institutions interacting more frequently and intensively with each other than with other persons and institutions.[3]

The evolution of the Hungarian S&T system will be characterized from the industrialization in the 1870s to the present time. Although the thesis will focus on the decades since World War II (WWII), the developments of the last 50 years cannot be explained without having understood the early industrialization of the country. At that time the S&T system was patterned after the Austrian and German systems, the two hegemonic powers of the region. While it is possible to discern a Hungarian science policy, it is difficult to identify a thoroughly planned technology policy from the 1870s to the late 1930s. Nevertheless, a number of institutions presently dominating the S&T system of the country were created at that time.

Therefore, one might speak of the establishment of the Hungarian S&T system in the 1870s. Yet, the changes in S&T were but a part of the rapid evolution of the Hungarian society during this period. The process of internal functional specialization sped up at that time. With the change from a feudal society to a mixed agrarian-industrial society an array of structures was created, which frequently featured previously non-existent functions, as for example economic governance institutions and a parliament. However, the time of industrialization does not in itself comprise a transition of the S&T system, since prior to industrialization there was no S&T system as such. Despite the existence of a number of institutions, like the Academy of Sciences or the university, these institutions led a rather isolated existence, not establishing close relations to each other so as to create a 'system' in the system theoretical sense.

Since WWII Hungary has gone through three transitions. During the late 1940s and early 1950s the S&T system, especially the industrial research and development (R&D) system, was transformed from a market based system to a centrally planned one. During this process the whole S&T system was compartmentalized and confronted with a centralization and bureaucratization of its control functions. It was the goal of this transformation to create a 1:1 copy of the Soviet S&T system, which was strongly influenced by a paradigmatic notion about the innovation process, which shall be referred to as the 'science push' policy paradigm.[4]

During the 1960s and 1970s large parts of the national economy, and with them the S&T system, went through a transition from a centrally planned and bureaucratically controlled to an indirectly bureaucratically controlled structure with integrated market elements. This study will provide indications that the Hungarian government reformed the S&T system using the one from the United States as a blueprint – without ever saying so. As early as in the 1960s, therefore Hungary utilized ideas stemming from a paradigmatic notion, here referred to as the 'demand pull paradigm', which was seriously discussed in the rest of Europe only after the Rothschild report induced reforms of the British S&T system in the early 1970s. The comparatively small Hungarian military-industrial complex was largely exempted from this development.

The third transition is still going on and is leading the country from the reform socialist economy back to a market based capitalist economy. The reforms of the years since 1989 have brought the country ever closer to a Western model. It will be shown that this time the S&T system is not only modeled after the United States, but also after the FRG. In doing so, the policy makers have imported new and more complex notions underlying S&T policies, based on the 'innovation process paradigm'. For a schematic comparison of the four time periods, the leading models of technological change and the major powers from which ideas on S&T policy were drawn, see table 1.1.

Table 1.1   S&T Policy Paradigms and Major Powers From Which Paradigms Were Drawn, 1870s –1990s

Time	S&T Policy Paradigms	Major Powers
1870s – 1940s	science push	Germany, Austria
1950s – 1960s	science push	USSR
1970s – 1980s	demand pull	USA
1980s – 1990s	innovation process	Germany, USA

Source: A number of papers and books, which have been influencing this categorization are listed in the annex.

At first sight the proliferation of these paradigmatic notions by the Soviet Union, the United States and the FRG might seem surprising. Yet, as will be shown in the study, paradigmatic notions underlying S&T policies as well as subsequent policy prototypes regularly emerge in large countries. Small countries regularly tend to apply these prefabricated paradigms and policies, indigenous paradigms and policies are comparatively weaker. However, a number of economic and social variables of each specific country constrain the idealtypical application of such paradigms and policies, even in smaller countries. Hungary will serve as an example for a small country; Germany, the United States and the Soviet Union for large powers. Austria will count as a large country, which lost its dominant position in the empire during the course of the second half of the 19th century and its status as large country in 1918.

While, in the course of the study, it will not be possible to prove the general hypothesis that small states mostly use paradigms and policies developed by large powers, a time series study of Hungary as an example of a small country with a rapid turnover of strongly differing political and economic systems will give strong indications for the validity of this hypothesis.

This book is structured in the following way: In chapter two, which directly follows the introduction, a number of the most influential theoretical models of technological change are discussed. Thereby the history of the ways of understanding innovation processes is shown, which is important for the historical analysis of the specific policies, the paradigmatic notions underlying the policies and the underlying notions' effects on the policies.

Then, a few key terms of systems theory are introduced and the relevance of Niklas Luhmann's theory for the analysis of S&T systems is explained. Furthermore, concepts based on system theoretical notions such as 'national innovation systems' or 'social systems of production' are introduced. Since the definitions of notions central to the study like 'technology' or 'science and technology system' are provided in the annex of the book, the main part of the study can then begin with chapter three.

In chapter three an abridgment of the development of the Hungarian S&T system since its origins is provided, with a detailed analysis of the system beginning with the industrialization which took place between 1867 and 1914. The time span between the two World Wars is only dealt with cursorily, because, due to the historic hardships the region suffered, this time was essentially a period of stagnation for large parts of the Hungarian S&T system. The country's S&T system is frequently compared with that of Austria, as the two countries formed the center of the Austro-Hungarian Empire, and Germany, because this country was economically furthest developed in the region. So Austria and Germany can serve as benchmarks for the success of Hungary's efforts in trying to catch up with the economically more developed part of Europe during this time.

Chapters three to six have an identical structure: on the basis of the historical background, which forms section one, the paradigmatic notions and policy goals are described in section two. Section three then analyzes the actual policies and their effects, whereas section four evaluates the policies and looks for possible explanations for the differences between policy goals and their outcomes.

In the fourth chapter the historical development of the Hungarian S&T system from 1945 until the late 1950s is briefly described against the background of the country's history. The year 1945 has been chosen as the starting point for a more thorough analysis of the S&T system because with WWII a new era of growth of S&T and governmental engagement in S&T began for the Western industrialized countries as well as for Hungary. Another reason for choosing this time frame was the regime change taking place in the country during the second half of the forties. First an abridgment of the reasons that led the Soviet Union to develop an S&T system with the structures it actually featured from the 1930s until the 1990s is given. Then a description of the changes the Hungarian S&T system went through during the late 1940s and the 1950s as a result of the copying of the Soviet S&T system is provided.

In chapter five the fate of the New Economic Mechanism, the reform program of the late 1960s' Kádár governments, is described. In both chapters four and five the intentions and the effects of the governmental policies and their underlying models are especially sharply contrasted. Reasons for the differences between theory and reality are considered: Why did neither realsocialist 'science-push' nor reform socialist 'science-pull' (or 'demand-pull') notions succeed in transforming the industrial R&D system?[5] Why were the – often quite substantial – changes in the Hungarian S&T system not successful in making especially the industrial R&D system as innovative as the systems of market economies?

In the sixth chapter I will take a look at the latest transition of Hungary from realsocialism to capitalism and the effects of these developments on the S&T system of Hungary. Thereby, a distinction between three periods will be made: the period of euphoria after the fall of the iron curtain, the phase of frustration, after the recognition that life would not necessarily get easier just through the introduction of capitalism, and the phase of realism, after the absorption of the new rules of capitalism throughout society, including the S&T system. In order to explain the differences between the expectations of policy makers and the actual effects of the policies employed, a group analysis of the Hungarian S&T system will be attempted and a patron-client scheme utilized during the last section of chapter six.

For reasons of stringency of argumentation, the comparative aspect of the Hungarian S&T system's development since the beginning of the latest transition will be discussed in chapter seven. In fact, the comparative aspect becomes dominant in this chapter, where Hungary will be compared with other countries with regards to three different aspects. First, the Hungarian national economy is compared with four Central European Free Trade Association Country (CEFTA) economies, namely Poland, the Czech Republic, Slovakia and Slovenia. Then, the Hungarian S&T system's in- and outputs are contrasted with similar figures of the CEFTA countries and Austria. Finally, the structures and functions of the S&T systems of Austria, Hungary and Slovenia are compared.

Finally, in the eighth chapter the research questions and hypotheses presented in the introduction are revisited and an effort is made to draw some lessons for S&T policy-making from the historical analyses of the Hungarian S&T system.

Notes

1   For a middle position, recognizing the complimentarity of both positions, see Brooks 1980 and Skolnikoff 1993.
2   For a definition of the term, see the annex.
3   A short definition of the term 'S&T system' is attempted in the annex, a short review of systems theory in chapter 2.2.
4   For a definition of the paradigmatic notions and ideas underlying the S&T policies in the time period studied, the three S&T 'policy paradigms' – science push, demand pull and innovation process – see the annex.
5   Science, considered the instrument for analyses of bourgeois and planning of socialist society by Marx, was termed 'a direct force of production' in the Program of the Communist Party of the Soviet Union at its 22nd Congress in 1961. This slogan, however, was the beginning of the official recognition of the needs to restructure the economic, and with it the S&T, system of the Council for Mutual Economic Assistance (COMECON) economies, by the Soviet Union itself. Hungary had at this point already engaged in this process of restructuring, despite the crackdown on it in 1956.

2 Theoretical Background

2.1 Theoretical Models of Technological Change

Since the time of England's industrial revolution which took place during the second half of the 18th century virtually no influential author has questioned the importance of technological change for a dynamically growing national economy. Nevertheless, there have been a number of different ways in which technological change has been understood. In this section a short look will be taken at the development of different models of technological change influencing people's conceptualization of economic growth, of the creation of wealth, or, more generally, of historical progress and development. These notions were not only interesting in themselves, but, as will be shown during the course of this study, also had an impact on how policies were conceptualized by governments, not only in Hungary, but around the World.

An understanding of the importance of the role of science and technology and of the progress of both can be traced back to the founders of modern economic thought during the Enlightenment of the 18th century. Adam Smith pointed out that there was a direct link between the ongoing division of labor and the improvements in machinery in his 'Wealth of Nations'.

> I shall only observe, therefore, that the invention of all those machines by which labour is so much facilitated and abridged seems to have been originally owing to the division of labour. (Smith 1974, p.114; orig. 1789)

Later Smith analyses the originators of the improvements in machinery, when he writes,

> (m)any improvements have been made by the ingenuity of the makers of the machines, when to make them became the business of a peculiar trade; and some by those who are called philosophers or man of speculation, whose trade it is not to do anything, but to observe everything; and who, upon that

16

account, are often capable of combining together the powers of the most distant and dissimilar objects. (Smith 1974, p.115)

Of course, Smith, writing in 1776, expressed himself differently than we would do that today. Nevertheless, his observations were right to the point. He observed that scientists ('philosophers') and engineers ('makers of the machines') were both responsible for the technological progress ('improvements in machinery'), which again was linked to the division of labor, enabling the development of machinery. Smith was not explicit about these problems, but it seems that although he saw technological progress as a driving force of development, he attached less importance to this factor than to the other classical factors of production, namely land, labor and capital, in his explanation for economic growth.

Karl Marx, for whose criticism of capitalism Smith's analyses were of fundamental importance, was not only among the first to identify all factors of production mentioned above, but he also placed a clearly larger emphasis on science and technology than did Smith. Under the impression of the industrialization of England, which was by then in full bloom, he saw technology as an important, if not the single most important driving force for historical change. In 'Die Grundrisse', written in 1857 and 1858, he related science, technology and production:

> [T]o the degree that large industry develops, the creation of real wealth comes to depend less on labour time and on the amount of labour employed than on the power of the agencies set in motion during labour time, whose 'powerful effectiveness' is itself in turn out of all proportion to the direct labour time spent on their production, but depends rather on the general state of science and on the progress of technology, or the application of this science to production. (The development of this science, especially natural science, and all others with the latter, is itself in turn related to the development of material production.) (Marx 1978, pp. 221; orig. 1857/58, p. 248)

In this passage Marx not only recognized the importance of large industries for the creation of wealth, he also stated that scientific and technological progress were more important to industrial growth than labor. In addition, he acknowledged the intertwined evolution of 'material production' and 'science'.

Lengthy discussions have been led on the topic of whether Marx perceived societal development, and therefore history, as being driven by technological progress or by social forces, such as the efforts of capitalists to increase their profit.[1] However, this discussion might be beside the point, as one could argue that both factors are interlinked and feedbacked and that Marx indeed related them to each other in this way. In any case, Marx's conceptions of the nature and relationship of science, technology and progress had a strong influence on how we perceive the processes of technological change and value creation today.

Just as Karl Marx drew from Adam Smith's work, regardless of ideological differences, so did Joseph Schumpeter from Marx. Schumpeter took Marx's notion of the disequilibrium of capitalist economies, in which the capitalists permanently strove to escape the 'law of falling returns' via innovative activities. This law can be characterized as an effect of the capitalists' efforts to engage into those economic sectors, which promised the largest profits. Since, however, all capitalists had similar information about the profitability of the individual sectors, most of them went into a small set of available highly profitable sectors at any given point in time. With competition rising and prices falling, the profits in these sectors markedly decreased. The only escape for the capitalists was to innovate and to create new products and processes in new economic sectors, which, until the other capitalists followed them into the new sectors, created short-lived monopoly situations and therefore promised high profit rates in the short term.

Schumpeter saw this as a process of 'creative destruction', in which old companies, which were not able or willing to innovate, died, whilst new ones, which were better able to adapt to the new environment and innovations were born. The often cited 'Schumpeterian entrepreneur'

> ...will, by [his] mere working and from within in the absence of all outside impulses or disturbances and even of 'growth' destroy any equilibrium that may have established itself or been in process of being established[.] (Schumpeter 1971, orig. 1928)

However, this entrepreneur was an integral part of 'competitive capitalism', which by the time Schumpeter wrote these lines, was already gradually being replaced by a new form of large-scale organizational entrepreneurship in the form of the new 'trustified capitalism', ultimately leading to 'an order of things which will be merely a matter of taste and terminology to call socialism or not'.[2] Capitalism would, Schumpeter was convinced, sooner or later become a victim of its own success and would find its heir in Communism.[3]

But more importantly, similar to Marx, Schumpeter was convinced that technological progress was the 'propelling force' of economic growth in general and of historical development in particular. Whatever the verdict on Schumpeter's estimations about the future of capitalism might be, he definitely influenced modern thought about science, technology, innovation and growth decisively through his dynamic view of innovation.

Yet in the 1930s and 1940s the time was not ripe for these aspects of Schumpeter's thoughts. Until the 1980s the standard economic view on growth, technological change and development was linear and static. The so-called neoclassical model, developed in this form by Robert Solow, included three factors of production, namely physical capital (subsuming land and capital), labor and technology. In this model the driving force for sustained economic growth was technological change. But since the model was not interested in technological change itself, technology was treated as an exogenous factor and fell under the 'ceteris paribus' (all other factors held equal) clause, which made it easier to model economic growth mathematically. Therefore, the model assumed that a constant rate of technological change existed, which provided innovations continuously and free of charge. Whilst the model was helpful in dealing with economic growth, it obviously was not informative with regard to technological change.[4]

In the 1960s a number of researchers started to become interested in technological change and produced results, which in time led to a revival of Schumpeter's ideas on growth, technological change and innovation. In 1962 Kenneth Arrow argued that the accumulation of knowledge in advanced industrialized economies was a by-product of mechanization, since machinery embodied knowledge just as much as it led to new innovations, i.e. new knowledge. In this perspective the process of knowledge generation assumes the form of 'learning-by-doing'. Moreover, Arrow proposed that, in order to increase the learning capacity of an economy, R&D – not only basic research – should be most efficiently carried out by an agency 'not governed by profit-and-loss criteria'. According to this view, R&D should be treated as public good, to be offered to firms at low cost, so as to counteract a possible underinvestment in R&D by companies.[5]

A number of empirical studies analyzing the production factors' input on growth,[6] which were carried out in search for new growth models which were to shed more light on technological change, a new set of models was born in the late 1970s. Schumpeter's vision of a dynamic and evolutionary economy was integrated into a number of studies (Nelson/Winter/Schuette 1976, pp.90–118; Nelson/Winter 1982), which transcended the disciplinary boundaries of economics and led to a view of economic growth and technological change, which has increasingly been rivaling the neoclassical economic model ever since.

The key difference between the old neoclassical models and the newer Schumpeterian ones is that the latter are more dynamic in their evolutionary perspectives. With regard to technological change this means an endogenization of the innovation process. Like the neoclassical model, the new models see technological change as the main driving factor for economic growth. However, since these models are interested in explaining technological change, they assume the production function to include factors such as the level of technology or more broadly the stock of knowledge, investments into R&D, skills of the work force (human capital), indicators of the complexity of institutional arrangements and the like, asides physical capital.

Moreover, the new Schumpeterian models can account for phenomena of technological change, such as spurs and slacks of certain technologies, since these models do not require a constant flow of exogenous technological change to ensure their functioning. On the contrary, these models thrive on a discontinuous rate of technological change, explaining innovations with the behavior of the archetypal 'Schumpeterian entrepreneur', who has been depicted above.[7]

With the rise of a new model for economic growth, which endogenized technological change, a different understanding of the innovation process itself permeated a number of disciplines, making itself felt also in the political sphere. Until the 1970s not only the notion of economic growth, but also the standard conception of innovation was linear. Incidentally, these ideas go back to classical social science theory. In the context of this study, it should be specifically interesting to analyze the ideas of Karl Marx on this topic.

Although it has been shown that Marx understood the intertwining of technology and science, he, like most social scientists until the second half of the 20th century, nevertheless believed in a model of innovation that often is referred to as the 'science-push' model. This idea is related to the notion of the linear technological 'pipeline', where basic science can be found at the beginning of the imaginary pipeline and the production of a good at its other end, with applied science, development and engineering, respectively, in-between. The term science-push arises from the idea that it is sufficient to increase the factor input on the side of basic science to get more competitive products as an output at the other end of the technological pipeline.

The underlying idea behind the classic Soviet S&T system implemented during Stalinism is that of the technological pipeline, thought to be driven by science. Nevertheless, decisive differences always distinguished this system from the Western applications of the science-push notion. The bureaucratization, centralization and compartmentalization, already mentioned above, were never shared to any comparable degree by the market economies. Moreover, even though Western advanced industrialized countries frequently spent a lower percentage of their GDP on R&D than did the realsocialist countries, their output in many scientific fields and most economics sectors was higher than that of the centrally planned economies.[8] The only exception to this might have been the time during WWII, when large capital flows were redirected towards R&D on both sides of the global confrontation between Allied Forces and Nazi-Germany and her allies in order to gain a technological edge. The whole understanding of the provision of public R&D was rapidly changing precisely because of the war.[9]

Jacob Schmookler was one of the first researchers to provide evidence that the science-push model was not an apt description of the innovation process. By analyzing almost a thousand cases of innovations in four different industries, he found that scientific discoveries were not the reason for any of the innovations he had looked at, but

> ...the recognition of a costly problem to be solved or a potentially profitable opportunity to be seized; in short, a technical problem or opportunity evaluated in economic terms. (Schmookler 1966, p.199, cited from: Grossman/Helpman 1991, p.5)

This clearer understanding of innovations which was driven by production and related activities such as engineering and development led to a new model of innovation. The 'science-push' model was turned on its head and was transformed into the 'science-pull' or 'demand-pull' model. The idea here is that the demand for innovations, i.e. the market forces, drive innovations and technological and scientific change.

In the Western market economies these notions had already found realization in the civilian R&D system of the United States, which was the most important role model for market economies in the post WWII era up till the 1970s. In the United Kingdom the Rothschild Report (Lord Rothschild 1971) aimed at reducing the inefficiencies of R&D paid for by the state via en bloc or institutional funding. Both ways of funneling money into R&D were found to lead to inefficiencies. The report proposed the introduction of the 'customer-contractor' relationship between publicly funded R&D institutions and their customers, among which were not only private organizations, but also governmental institutions. This 'simulated market' was supposed to introduce a new, more efficient way of working within the R&D institutions by forcing the previously fully state funded organizations to react to demand.

Very similar ideas can be found in the reforms which several centrally planned economies introduced during the 1960s.[10] Arguably, the approach most visibly breaking with the traditions of the classic Soviet system – at least in its original intentions – was undertaken with the Hungarian reform-program, the 'New Economic Mechanism' in 1968. The principles employed in the S&T system, but even more so in the industrial R&D system were similar to those in the Rothschild report.[11] In Hungary en bloc funding of institutions by government had already been reduced during the first half of the 1960s. Since 1968 R&D institutions were encouraged to contract out their services for companies and other organizations, governmental and non-governmental.

In the light of the unsatisfactory results gained through the policies employed in both market economies and centrally planned economies the notion of the science- or demand-pull model was modified during the second half of the 1970s.[12] The new model which stood behind science, technology and innovation policies has been termed 'complex model' by Beatriz Ruivo (Ruivo 1994, pp. 157). Amongst other things, as a result of the rise of the new Schumpeterian models of growth, in this model the understanding of technological change is not linear any more. The relationship between each part of what was previously imagined as the technological pipeline is now understood as being highly complex. Richard Nelson and Nathan Rosenberg point out that 'saying that new technologies have given rise to new sciences is at least as true as the other way around' (Nelson/Rosenberg 1993, p. 7).

Eugene Skolnikoff sums up today's thinking about the relationship between science and technology:

> The relation between science and technology is complex, contains many feedback paths, and cannot be characterized by a simple linear progression from science to technology to application, even though a higher proportion of technology is science-based today than in earlier eras. (Skolnikoff 1993, p.15)

For the increasing science-dependency of technology over the last years Skolnikoff gives examples from fields as diverse as molecular biology, physics of materials and mathematics, which are likely to have effects on agricultural productivity, superconductivity and computer design.

Niklas Luhmann, who analyses the relationship of science and technology from an entirely different point of view than Skolnikoff, comes to a similar conclusion all the same.

> Es liegt auf der Hand und wird wohl kaum bestritten werden, daß die Technologie in zunehmendem Maße wissenschaftsabhängig geworden ist, so wie die Forschung selbst technologieabhängig. (Luhmann 1990, p.264)

Here Luhmann comes to the conclusion that, increasingly, science depends as much on technology as technology does on science. Later in the same text Luhmann insists that science and technology cannot be divided in a dichotomy of 'theory vs. practice'. If anything, technology is a contribution of science to society, but not the function of science. In turn, he writes, technology influences science in a fundamental way:

> Mit all dem spielt 'Konstruktion' und genaue Analyse des 'Wie' der erkennenden Operationen eine viel größere Rolle als früher und ersetzt mehr und mehr die Frage nach dem 'Was' der Erkenntnis. Die Steigerungslinien, die sich gegenwärtig abzeichnen und die Wissenschaft und Technologie immer enger zusammenschließen, entsprechen genau diesem Muster. (Luhmann 1990, p.265)

In this quotation Luhmann points out that the question as to 'how' increasingly replaces the question as to 'what' in science, that construction and the knowledge about it gains importance over the cognition. He sees this development as being in line with the general trend of an increasing dependency of science and technology on each other.

Christopher Freeman tries to reconcile Schumpeter's and Schmookler's work by finding that demand-led models of innovation do not match his own findings for a variety of industries.

> Schumpeter's theory of an autonomous impetus on the supply side, deriving from advances in science and invention and realized through imaginative entrepreneurship, appears to fit the facts rather better. As we have seen, however, once a major innovation has been made, then a pattern of demand-led secondary inventions and innovations may set in over many decades giving apparent credibility to a 'Schmookler'-type of analysis. (Freeman 1982, p.211)

At present there is no single clearly identifiable and easily comprehensible model at hand aiming at a description of the process of technological change, but rather a multitude of models describing facets of the process.[13] This also has effects on policy makers. S&T policies today have to be rather more pragmatic than dogmatic, utilizing a wide variety of policy instruments rather than concentrating on a single policy.

2.2 Systems Theory and S&T Systems

Modern societies feature an ever increasing degree of complexity. Many functions, which in previous time periods were performed by a single organization or even a single person, are nowadays tasks fulfilled by a variety of different institutions. Just as in the area of science and technology (S&T) the university was the dominant institution for hundreds of years, in the area of the economy the single company was highly important for an equally long period of time. Yet, in the course of the 20th century the university transformed its structures and functions and now, although it is still an important producer and user of knowledge, it has become one organization amongst many to engage in S&T, together with a variety of public and private research institutions. Similarly, the single firm, which still is the most important economic unit in advanced industrialized economies, has changed its structures and functions massively and can now take on a number of different forms, being integrated in a variety of private and public networks.

For a comprehensive understanding of these transformations and the increasing complexity of societal structures the repeatedly mentioned notion of 'systems' is helpful. By utilizing the idea of systems one can draw boundaries between different societal spheres. Thereby the relations of these spheres can be described in order to achieve a better understanding of S&T itself.

As might be inferred from the term, the notion of 'S&T systems' stems from systems theory. Not only can an analysis of societal complexity and technological change be developed in the framework of systems theory, but going even further the set of tools, which systems theoreticians have developed over time, is helpful for a new understanding of the relations of S&T to other spheres of society. But before the virtues of systems theory for the description and analysis of S&T systems are elaborated in detail, a brief look shall be taken at theories of complex social systems, and a closer look at the kind of systems theories directly applicable to the S&T systems, and, in particular, Niklas Luhmann's theory.

Systems theory stems from at least two strains of thought. One is the drive of scientists to arrive from a pure analysis at a synthesis. This effort has a considerable history and includes a wide range of original dialectic thinkers from Socrates to Karl Marx as well as other traditions, for example the functionalists Talcott Parsons and David Easton. Anatol Rapoport points out that the distinction between analytically and synthetically holistic thinkers cannot only be found in the social sciences. He provides examples ranging from mathematics to psychology and from medieval ontology to economics (Rapoport 1992, pp.5). The other strain of thought influencing systems theory is the drive of science to compare systems, often in the form of models. This effort has a comparably long history including thinkers as diverse as Niccolò Macchiavelli, Max Weber and Immanuel Wallerstein.

Modern systems theory distinguishes between a number of approaches. They originate in functional theories, which were dominating social sciences in the 1960s and 1970s. Functionalism can be described as the effort to find specific – and detrimental – mechanisms that are crucial to the regulation of the workings of a body, institution or society. This approach is linked with scholars like David Easton, Gabriel Almond and Talcott Parsons, who during the 1950s and 1960s applied systems theories to the social sciences.

Helmut Willke distinguishes a number of approaches, which are all systems theories. He examines

– the structural-functional approach,

which was the first systems theory of Parsons, in which structures are seen as an independent variable and only functions are to be explained;

— the system-functional approach,
   which concentrates on internal functions of the system and thereby largely ignores the environment of a system, but comes a step closer to understanding the structure of a system as a dependent variable;

— the functional-structural approach,
   as developed especially by Niklas Luhmann, which concentrates on the relationship between system and environment, under the presumption that the reason for the creation of a system is found in the reduction of complexity of the modern world, with the structures of a system being built in reaction to these functional needs;

— the functional-genetic approach,
   which concentrates even more on the processes and especially the evolutionary developments of systems; and

— the self-referential approach,
   which centers around the notion of autopoiesis, i.e. the idea that complex systems reproduce themselves continuously out of the elements they are constituted from. Luhmann incorporated the theory of autopoiesis, which was developed by Humberto Maturana and Francisco Varela, into his own work (Luhmann 1984, pp.60) under the heading 'self-referentiality' (Willke 1987, pp.3).

Nowadays, the arguably most influential systems theory addressing complex social systems is the one created by Niklas Luhmann. The theory is used explicitly in fields as diverse as business administration,[14] political science and sociology of science.[15] The theory not only has characteristics making it compatible to the analysis of S&T systems, which are going to be elaborated on later, but is also well formulated by now.[16] In the following paragraphs an effort will be made to characterize Luhmann's theory, without simplifying it too much – a difficult task considering the complexity of the theory at hand.

An overview of Luhmann's theory should start by explaining the basis of his analysis of complex social systems, which is society itself. For Luhmann society includes everything social, i.e. all communications and activities of the elements inside society. The term 'communication' is crucial here, as it is the mechanism reproducing society.

Since the wake of industrialization, the complexity within societies has been rising sharply. In an effort to cope with this complexity structures have arisen, which create a functional differentiation within society. An example might be a village of farmers, which for generations individually milled their own corn and baked their own bread. Due to rising population and production levels, the community is capable of allowing a single family to specialize in corn milling, whose living can be sustained because enough corn is produced in the surroundings of the village to keep the miller at work all the year round. After another generation a baker might find enough demand to open up a business. With other farmers specializing in a single occupation, the local villagers slowly differentiate their societal functions in the sense of divisions of labor occurring more and more continuously. The new occupations and structures could not be found in older societies precisely because previous societies were lacking the need for the functions carried out by these smaller units, be it millers, bakers, blacksmiths, mechanicians, doctors or judges (compare with Luhmann 1990, pp. 298). The people in new occupations will find that they work together more or less frequently, depending on their specializations. The policeman might work more closely together with the lawyer and the judge than with the blacksmith and the miller.

Over time, with ever more increasing specialization levels, structures will be formed, which can be understood as 'subsystems', analogous to society, which is called 'system'. An example for such a subsystem might be the local police. After the village, which needed only a single policeman, has grown into a town, a single policeman might have been joined by a number of colleagues, which will have to be placed in several police-stations, spread over the area of the community. In a system theoretical view, this network of closely cooperating police-stations, which again might differentiate into police units patrolling the streets, others guarding objects and plain cloth units fighting organized crime, might be defined as a subsystem of society.

The definition of systems utilized for such cases is quite straight-forward: Systems are elements featuring interactions and relations relatively more intensive amongst each other than to other elements. For our purposes the usage of the term systems stands for 'complex systems', in the language of cybernetics 'non-trivial machines', a concept developed by Heinz Foerster.[17] In the case of 'trivial machines' the input of an environment is transformed into a specifiable output, which stays the same, as long as the input is not changed, regardless of the machine's number of work-cycles. Therefore, as soon as the function of the trivial machine is known, the output is known, if the input is known. Conversely, the input is known, if the output is known. In the case of non-trivial machines the input is transformed into an output, too. However, it is not possible to know the output, even if the input is known, since the function of a non-trivial machine is not determinable. In a comparatively simple case the non-trivial machine consists of a trivial machine inside a trivial machine. David Easton used a function creating a 'withinput' between input and output for similar purposes (Easton 1965).

The crucial element for a system is the 'borderline' between the system and its environment. The borderline is a function of the mechanisms dividing interactions and communications into those belonging to a certain system and those which do not belong to the system. The prime mechanism of the crucial borderline maintenance ('Grenzerhaltung') is the allocation of meaning ('Sinn'). Meaning, essential to the lives of all humans, is present in the form of, for instance, roles, values, and 'Weltanschauungen' – and it is created continuously through interactions.[18]

At this point it has to be emphasized that systems are open in the sense that they interact with their environment – across the borderlines described above. Furthermore, systems can only be understood in their relationship to their environment and in their state of being different from their environment.[19] There is only one exception to this general statement, which addresses the largest complex social system – society itself. As there is nothing social outside society, it has to be closed by definition. All other complex social systems are subsystems of society and have to interact with their environment.

Everything that has been said about systems is also valid for subsystems. Subsystems are not passive lower order units of the larger systems. They autonomously function in the sense of their operational closure. However, they are not closed in the sense of causal autonomy. As has been indicated before, they are indirectly dependent on their environment – a statement that will become clearer after a short look at the concepts of autopoiesis and self-referentiality.

Complex social systems are characterized by 'autopoiesis', which means that only the system itself can reproduce itself. Systems are also 'self-referential' in this sense. A system's essential autopoietic functions are key to the survival of the system in its environment. The self-reproduction of a system is related to its environment and even creates the possibility of establishing relations to the environment for the system. Therefore self-referentiality is also the system's reaction to its environment. Consequently, the system can be seen and interpreted only via the system's reaction to its environment (Luhmann 1984, pp.57).

Another term central to Luhmann's theory is the 'media'. 'Media' is not used in the sense of instruments of mass media. Luhmann in this context also uses the term 'symbolically generalized steering media' ('symbolische generalisierte Steuerungsmedien') to signify the differences between the term 'medium' in everyday language and in his theory. 'Media' are general code-structures proliferating meaning specific to a certain (sub)system. Examples for media in complex social systems are power (for the subsystem politics), money (for the subsystem economy), knowledge (for the subsystem science) or faith (for the subsystem religion). Media provide sets of contingencies for transactions inside a system. In this sense they structure the communication between the elements of a system.[20]

Willke provides an example for the evolution of a specific medium: money. Media with the function of money were developed in societies relatively early. The role of the medium was to make transactions easier and to enhance the steering functions vis-à-vis the economy. After a time the economic transactions became less and less dependent on specific persons, but, as an effect of rising societal complexity, became a function of institutions. Thereby, first money, and later capital as such, became the very center of the economy.

This process becomes visible in such environments as the stock market, where virtual capital is traded in various forms, including such high-risk instruments as options. These instruments are often not used with the explicit goal of an actual realization of the invested capital, i.e. they are not exchanged into tangibles of some sorts, but are reinvested. In fact, the international stock market's operation is contingent on the fact that the traded stocks are not realized, since this would exceed the monetary possibilities of the global financial system.

Then again, the economic effects of the world-wide stock market crash in 1987 and the burst of the Japanese bubble economy in the early 1990s, which again was based on virtual capital, this time in the form of foul credits, have shown the double nature of the medium money. Due to the rising trade volumes of the international stock markets, the effects of the virtual stock market capital are quite manifest and not virtual at all. It is in this sense that the medium money has increasingly gained a life of its own, formulating its own specific set of rules. These rules have been changing the conditions of the transactions. The medium money facilitates transactions through a new level of abstraction and generalization. As a result the criteria for decisions in the economic system are now heavily influenced by the medium money (compare with Willke 1987, p.137).

A number of problems in the analysis of S&T systems can be handled with the help of Luhmann's theory in a rather convenient fashion. One of the reasons for the analysis of S&T systems is the drive to maximize the output of these systems under scarce resources, to put the problem into the terms of neoclassical economics. One of the reasons why this is so hard is the incompatibility between the different involved systems. The economic, the political, the scientific and the technological systems work according to different logics and linkages between them often are rather unsuccessful. Systems theory provides helpful tools for an analysis of the situation.

A good explanation for the linkage problems so central to the structures of S&T systems uses Luhmann's 'media'. As has been mentioned, different subsystems utilize different media for steering purposes. The internal logics of these media differ to a large degree. Here, a good example is provided by those two systems, whose relationship is perhaps the most important to a S&T system: science and economy. While the economy uses the above-mentioned medium 'money', the medium in science, according to Luhmann, is 'truth' ('Wahrheit').[21] The internal logics of the two systems based on these two entirely different media are largely incompatible due to the different value structures and resulting behaviors of the subjects which are part of the systems. A possible solution might be the rise of 'supermedia', which influence more than one subsystem – an example of which might be money, if one accepts the dominance of the economic system as a result of globalization and other recent developments (Willke 1987, pp.147). Another solution might be the rise of intermediary organizations or transformation enhancing structures as for example corporatist institutions, in which partners from the economy, politics and science might be included.[22]

The elegance of Luhmann's theory lies in the fact that it can exist without presuppositions, which are necessary for neo-classical economic theory or liberal theories in general: no ideal information situation, no rational behavior of humans is necessary, no maximization of profits inherent to their doings. Marxism's class struggle, the disequilibria of all societal units, is not called for, either. The replacement of the fixed basic unit of an analysis of liberalism (the individual), of Marxism (the class) or of statism/nationalism (the nation) through the comparatively flexible notion of systems is helpful in many cases, too. All this frees the analysis from restraints but, doubtlessly, creates new restrictions.

In any case, systems theory cannot replace liberal or Marxist thought. On to the contrary, the limits of systems theory should be recognized and each theoretical frame, be it liberal, Marxist, statist, systems or other theories, should be applied to problems fit for the respective approaches. On the one hand, it might be difficult to criticize widening income gaps with systems theory, but it is easy to find reasons for such a phenomenon with the instruments of Marxism. On the other hand, it might be difficult to analyze the technology transfer problems in S&T systems with Marxist conceptions of organizations, but easy to do so with systems theory, as has been shown in the case of incompatibilities of subsystems due to their usage of different media.

Moreover, despite the possibilities that an analysis of S&T systems with the tools of systems theory offers, distinct weak points of Luhmann's theory have to be acknowledged. The perhaps most persistent criticism brought forward against Luhmann is that he does not systematically address the problem of the borders of systems. In the real world, according to the critics, the functional differentiation of different subsystems does not draw clear-cut boundaries. The charges against Luhmann's theory center around the blurring of functions, especially between the subsystems of the economy and politics. If a politician makes a decision about economic policy, does he act in the subsystem politics or the subsystem economics? If a citizen pays a doctor for a diagnosis, is he in the subsystem economics or the subsystem health services? Luhmann reacted to these criticisms first with long replies in his books, in which he defined functions and variables of the concept, later with rather short stereotypical answers.[23]

A second line of criticism aims at the missing role of the individual in Luhmann's theory. The individual as such, according to the critics, is not discussed by Luhmann. Wolf Krohn and Günther Küppers have been persistently critical of Luhmann in this respect and, as a result, have developed their own theory of self-organization of science, in which they center on the individual researcher as their basic unit of analysis. In addition, Krohn and Küppers do not insist on the strict boundaries of the respective subsystems, which allows for changeable borders and even hybrid areas, in which systems can interact and establish new forms of communication.[24]

Luhmann has also been charged for his lack of interest in technology. Technology for Luhmann is not a subsystem of society, but an intersystemic structure located 'between' the subsystems. This gives a lot of flexibility to the notion of technology as an output and input variable to a number of subsystems. However, Luhmann has not systematically elaborated this concept further.[25]

A more fundamental criticism of Luhmann's theory is the charge that the practicability of his theoretical concepts is low. Not only the two areas, the boundaries of systems and the role of the individual, but also the general scope of Luhmann's work, which often is highly theoretical, support this argument. As has been pointed out before, Luhmann, in contrast to other systems theorists like Helmut Willke, was not always inclined to apply his theory to reality. Luhmann himself has always replied to attacks at the complexity of his theory with the remark that other theories are under-complex and therefore cannot be seen as good models of a complex society.

These and other criticisms against Luhmann's theory are certainly correct. However, the merit of Luhmann's systems theory, as it has been discussed before, seems to outlast the criticism. The flexibility of the concept, the non-usage of presuppositions and the ability to explain the different logic of subsystems are arguments for the usage of Luhmann's theory to the end of the analysis of S&T systems. Nevertheless, the shortcomings of Luhmann's construction should be kept in mind and there should be no hesitation to fill in gaps left by Luhmann with other concepts, both older and newer ones.

And indeed, since the 1980s a number of concepts, which are based on the systems notion, have been used for the understanding of technological change and the relationship of S&T with society. These concepts share a number of underlying ideas with Luhmann's system theory, but are more flexible and much less strict with respect to their academic rigor. One example of a system concept is the technological system approach, developed in the early 1980s, by theorists like Thomas Hughes (Hughes 1983, 1994). In the late 1980s the notion of the national systems of innovation was developed by Freeman, Lundvall, Nelson and others Freeman 1987, Lundvall 1992, Nelson 1993). A few years later the term knowledge systems was coined to describe essentially the same idea, but with a different analytical focus (Smith 1994). At the heart of the knowledge system notion lies the distribution power of the system, primarily the distribution of knowledge between universities, research institutions and industry (David/Foray 1994).

The most widely used systemic concept of the 1990s in this context is undoubtedly the notion of the national systems of innovation. The first definition of a national system of innovation, which over time has proven to be robust, was proposed by Freeman in 1987: 'The network of institutions in the public and the private sectors whose activities and interactions initiate, import, modify and diffuse new technologies may be described as the "national system of innovation"'.

Despite the conciseness of the wording, the quotation brings to the open the same problem as featured by all definitions for national systems of innovation. In an effort to close the system, but still retain the most important institutions and interactions, its borders are widened and numbers and forms of included organizations grow almost exponentially. This has been proven again and again in efforts to pin down national system of innovation empirically (see for example Nelson 1993).

Another solution to this problem has been chosen by Rogers Hollingsworth and Robert Boyer (1997), who developed the concept of 'social systems of production'. While the focus of this notion does not lie in the innovative activities of a system, these activities play an important role as a mechanism of change in the model. In order to evade the pitfall of precluding the analysis to the national level – a problem, which has led to serious criticism of the national systems of innovation concept – social systems of production have flexible boundaries, which do not necessarily have to be confined to the nation state.

Boyer and Hollingsworth (1997, p.2) define social systems of production as

> the way that the following institutions or structures of a country or a region are integrated into a social configuration: the industrial relations system; the system of training of workers and managers; the internal structure of corporate firms; the structured relationships among firms in the same industry on the one hand, and on the other firms' relationships with their suppliers and customers; the financial markets of a society; the concepts of fairness and justice held by capital and labor; the structure of the state and its policies; and

a society's idiosyncratic customs and traditions as well as norms, moral principles, rules, laws and recipes for action.

What is interesting about the social systems of production is that they consist of institutions, which conform to rules and are themselves are based upon norms and values. Therefore the concept elegantly explains differences in structures and working mechanisms of national economies as a result of the variation of the cultures the systems are embedded in and contingent on. The mechanism causing the structures and functions of a social system of production to conform to the basic rules of a society are the system's ordering principles, which can be the market, the hierarchy, the community, the state, the network and the association (Streeck/Schmitter 1985, Hollingsworth/Boyer 1997). In most advanced industrialized societies all of these principles are represented, with some being predominant vis-à-vis the others, however. Depending on the relative strength of each of these ordering principles, social systems of production are more likely to feature structures and working mechanisms as Germany, where the association and the hierarchy are the dominant ordering principles, the United States, where the market and the network are dominant, or Japan, where the community and the hierarchy are dominant.

The concept is helpful in a number of respects: First, it is an additional and, with respect to systems theory, sometimes indeed alternative explanation for the problems in the interaction between the various societal subsystems. Within the framework of the social system of production, the logic of the specific subsystems can be explained by the variances in institutionally based norms and rules[26] which are also reflected in the strength of specific ordering principles varying between institutions as well as societies. Thereby, the differences in norms and values, which constitute and embed institutions and on which the relative strength of the ordering principles is contingent are an explanation for the workings of an institution – making this concept an alternative to the analysis centering on the above mentioned symbolically generalized steering media, which stems from systems theory. A notable difference between the two explanations is that the social systems of production notion stresses the interaction between the institutions more than the functional-structural systems theory as developed by Niklas Luhmann does. This creates a certain supplementarity between the two notions, which shall therefore both be kept in mind for what is to follow.

Another facet of the notion of social systems of production will be studied more closely here. Since the concept is based upon culture in the form of norms, values and rules advancing certain ordering principles and restraining others, it appears to be rather static, because culture is rarely subject to quick changes, but transforms only gradually. Therefore, the social systems of production notion can be put to a test in the course of this study: if indeed the concept is right in basing the institutions of a country on culture, it can be predicted that the social system of production as such, but moreover also the institutions it consists of, shall be subject to only slow and gradual change (Hollingsworth 1998).

The notion of social systems of production can be used to describe and analyze a possible resistance to change of institutions or whole societies, which frequently can be found in the study of S&T (Landes 1969, Bijker 1995, Biegelbauer 1997, Hage 1998). Such rigidities can take several forms: on the one hand, it can be the effect of a path dependency in which a certain institution or group of institutions at a specific point in time may have set a certain course of action. This decision might be driven endogenously, as for example by a certain technology, which has been chosen by the institution and has become more and more important for its functioning. It might also be created exogenously, as through the differentiation of a certain function by society, which the institution has to fulfill, or the development of a specific mission, which the institution is expected to follow.

On the other hand, the resistance to change can be unspecific and general, in this case it can be the result of a number of factors. Again, the institutional rigidity can be generated by internal processes, as for example a certain constellation of interests within the institution, which produces a deadlock situation. It can be exogenously produced, too; in this case an institution, which is hierarchically higher than the former institution, might hinder it from changing. Another possibility might be that an institution determined to change cannot do so, simply because it is interlinked with a number of other institutions, for example as a result of production relations, which do not want to or cannot change themselves and indirectly and perhaps even unwillingly hinder the former institution to change.

Yet another reason why the social systems of production notion is helpful lies in its flexibility. Similarly to systems theory, which has been utilized here for the conceptualization of society, the integrity of the social systems of production concept is maintained, even if the notion is supplemented by other approaches for a better understanding of societal phenomena. Since neither systems theory nor the social systems of production notion are helpful for the analysis of interest formations in society, group theory, together with a patron-client scheme, shall be used for this purpose. More specifically, a group analysis (Bentley 1909, Truman 1956, Hrebenar/Scott 1990) of the Hungarian S&T sphere will be carried out as a part of the evaluation of the Hungarian S&T policies during the latest transition from realsocialism to capitalism.

Although the notion of social systems of production is a helpful one, society shall be classified in the very system-theoretical terms already laid out before. The reason for this decision lies in the fact that the social systems of production concept makes it difficult to focus on a single societal subsystem (or 'institutional sector', in Hollingsworth's (1998) terms). Moreover, single components of a social system of production may be parts of several subsystems of society and can be separated from these subsystems only with great difficulty. Boundary definitions are unproblematic for a number of analytical observations, as the ones, which the social systems of production notion has been using, but it is a problem for a time-series study of a singular subsystem of society, such as the one at hand.

In an effort to escape this dilemma, a core area of the 'national system of innovation' or 'social system of production' shall be proposed, namely the 'science and technology (S&T) system'. The S&T system describes the central part of a national system of innovation responsible for the larger part of the initiation, import, modification and diffusion of technology. The analysis of the S&T system should, therefore, focus on the central research, development and engineering (R, D&E) as well as educational institutions of a national economy and the political structures directly related to these organizations, i.e. councils, ministries and agencies that supervise the institutions listed above.[27] While this definition of what an analysis of a S&T system should entail might seem a bit crude, it should be clear-cut enough to minimize the boundary problem and should therefore be kept unchanged over the period of analysis.

A number of indicators will be employed in the following analysis of the Hungarian S&T system. Three sets of factors of outstanding impact on the S&T system will be used in particular as indicators for the analysis of the evolution of the system as well as of the policies employed by government to reform the system, namely:

- the allocation of resources, especially the nature and scale of expenditures on S&T as well as the number and skill level of people engaged in S&T,
- the institutional structures and functions, particularly those organizations responsible for the planning and financing of as well as engaging in S&T,
- the social system of production, with a special emphasis on the S&T system, for instance the output of the S&T system itself, but, extending the S&T system, also the structure of macro demand and the in- and output of industry at large.

While there is an emphasis on the first two sets of indicators in the chapters dealing with the earlier transitions of the Hungarian S&T system, the description and analysis of the latest transition, beginning in the late 1980s, rests stronger on the last group of indicators. The reason for this lies in the fact that in the last chapter the policy evaluation aspect is somewhat stronger and therefore the last set of indicators is more frequently used in comparison of the economies and S&T systems of several CEECs.

Notes

1   For a reflection of this argument, see Robert Heilbroner's classical article 'Do Machines Make History'? (in Smith/Marx 1994, pp.53–66), where he advances the technology argument; in the same volume, Bruce Bimber favors the societal argument (pp.79–100).

2   Schumpeter 1971, p. 42; compare also with Schumpeter's later major work 'Capitalism, Socialism and Democracy' (1975, orig. 1942); here he elaborated on motives one can find in his writings throughout his whole life. It is interesting that today's Schumpeterian economists, whose work is increasingly recognized in areas such as growth theory, innovation theory and neighbouring fields, seldom mention these parts of Schumpeterian thoughts. The renowned economist F.M. Scherer, for example, does not mention any linkage between Marx and Schumpeter – except for pointing out that 'economists of as diverse persuasion as Adam Smith, Karl Marx, and Joseph Schumpeter have argued that material standards of living depend critically on the level of technology' (Scherer 1984, p. 32).

3   There was a lively discussion during the second half of the 1980s as to whether the economic systems in the OECD countries should still be termed capitalistic or whether they were already a form of socialism. See for example Gorz 1985 and McCormack 1990, pp.1.

4   For more information on the neoclassical model of economic growth and its ramifications for the understanding of technological change see Valdés 1999, pp.15.

5   For an account of this idea, which is surprising at first sight, see Rosenberg 1971 (orig. 1962).

6   For a comparison, see Boskin/Lau 1992, p.32.

7   Compare with Nelson 1992, pp.57.

8   Frequently data from international comparisons of the Organization for Economic Cooperation and Development (OECD), the US Central Intelligence Agency (CIA) and the US National Science Foundation (NSF) showed higher percentages of the GDPs of centrally planned economies spent on R&D than was the case in Western industrialized countries. See for example Hanson/Pavitt 1987, p.53.

9   Not only were the factor inputs raised rapidly, but also was the organization of state funded R&D reformed. For the history of the US Office of Scientific Research and Development (OSRD) see Stewart 1948. For a history of the beginning years of the US National Science Foundation (NSF) see England 1982. Both organizations bore the imprints of Vannevar Bush, who also was the author of the now famous report 'Science – The Endless Frontier', advocating public engagement in the field of basic research. The NSF has been the leading role model for the organization of basic research during the whole post WWII period for market economies. Nowadays the institution plays a similar role for the formerly centrally planned economies. See also Skolnikoff 1993, pp.16.

10   For an overview of the efforts of the GDR, Hungary and Yugoslavia see Keren 1992.

11   For an appraisal of this fact see Balázs, 1993.

12   Among the OECD countries especially the UK carried the notion of the demand- or science-pull strategy furthest – only to have to turn to a mix between market-driven and publicly funded S&T strategies in the 1970s. Since the implementation of the NEM, Hungary was the centrally planned economy adhering at least in theory most closely to the demand-pull notion. Both countries' experiences seem to suggest that a complete reliance on one of the two models is not successful.

13   Coming closest to a standard model is the concept, which has been provided by Kline and Rosenberg (1986).

14   For instance he is cited at length in one of the standard German language text books on organization and strategic firm leadership, Schertler 1988, pp.246.

15   In both fields Luhmann's theories are not only widely cited, but his thinking has influenced the general conceptualization of problems. This is all the more true since it was not viable to replace the three large *Weltanschauungen* of

Marxism, Liberalism and Statism/Etatism/Nationalism with theories of comparable range. Systems theory, though lacking a teleology and therefore being rather a mid-range theory, since the 1980s has been one of the possible successors for the Weltanschauungen.

16 This should be especially emphasized since Luhmann developed his theory over a period of more than three decades, sticking to his basic ideas, but reformulating and refining them until his death in 1998.

17 For a brief depiction of the differences between trivial and non-trivial machines and the effects of this theorizing on systems and organizational theory, see, Schimank 1987, pp.45–64, or, for the original author's statements, Foerster 1984, pp.9.

18 For a longer definition of this notion central to systems theories – the border between systems and their environment – see Luhmann 1984, pp.35, for a shorter version, see Willke 1987, p.175.

19 As has been said, it is precisely this difference that makes systems possible. If there is no difference between system and environment, there is no system. See, for example, Luhmann 1984, pp. 22, Willke 1987, p. 93.

20 Luhmann 1984, pp. 222; for the development of the concept since the 1950s, see Willke 1987, pp. 138.

21 For Luhmann's concept of 'Wahrheit' ('truth'), see for example Luhmann 1990, p. 198; although it is not entirely clear how close 'Wahrheit' in Luhmann's sense comes to the Anglo-Saxon 'correctness', 'validity' or really to 'truth' (with all its dramatic and almost religious undertones), it might not be necessary to analyze the term in depth. After all, Luhmann has an entirely constructivist understanding of the term Wahrheit – see, for example Luhmann 1990, pp. 70 and p. 701. On page 198 of his 1990 volume he writes that Wahrheit is constituted by a system and that if something is seen as true, there is a system-internal value judgement behind the term, which is used as a symbolically generalized medium.

22 Luhmann has not been very interested in these questions. Willke points out that both Parsons and Luhmann are stating the problem, without analyzing possible solutions. Willke himself has made efforts to use systems theory for practical purposes again and again. I have met Helmut Willke for the first time at a workshop in 1991, where Willke used systems theory to explain the current problems of the SPÖ as a modern catch-all party increasingly failing to cope with the growing functionalization of modern society.

23 In 1996 I attended a conference in Bielefeld, where Niklas Luhmann took part in a panel discussion. When Luhmann was confronted with the weaknesses of his concepts of systems' borders in real life, he answered several times, almost like a mantra, that these were not actual problems and that the functional differentiation of systems allowed for a precise distinction.

24 For a short description of some of the key differences between Krohn/Küppers' and Luhmann's notions of systems, see Weyer 1989, pp. 101.

25 Weyer (1989, page 100) points at the underdefined areas of Luhmann's intersystemic relations and mentions technology as such an underdefined area, which, however, could serve as the material to fill the void between the subsystems. Moreover, Weyer sees the possibility to break through the comparatively narrow limitations of intersystemic relations in technology serving as a medium between the subsystems.

26 A notion based upon neo-institutionalism, as developed by Douglass North (1981, 1990).

27 For a more detailed definition, see the Annex.

3 The Foundations:
Establishing a National Science and Technology System

3.1 Historical Development of the S&T System

Hungary is a Gerschenkronian 'late late developer'[1] in most aspects of its economic and technological development (Gerschenkron 1962, 1977). This statement, however, does not hold true for the area of science,[2] where Hungarians have shown their capacity to engage into quality research at least since the beginning of the industrial revolution at the end of the 19th century.[3]

Due to its economic, cultural and historical connection to Western and Central Europe, the establishment of the first universities in Hungary is in line with the founding of other similar institutions in Europe. Universities were founded in 1367 in Pecs, 1395 in Ó-Buda and in 1467 in Pozsony, today's Bratislava. However, these institutions never existed for more than one or two generations for neither was urban life advanced nor royal support steady enough to make such an institution possible (Tamás 1985, p.31). Hungarian students regularly went abroad, to Germany, Italy or other intellectual centers, in their effort to gain knowledge. Indigenous Hungarian intellectual life diminished during the 16th and 17th century, when the country became part of the Ottoman Empire.

During the time of the wars with the Ottoman Empire Hungary remained on the periphery of Europe. However, its integration into the empire of the 'Habsburgs' did not change the status of the country. Nevertheless, in 1635 Cardinal Pèter Pàzmàny founded the first Hungarian university still in existence today (nowadays called Eötvös Lorànd University) in Tyrnau. In 1777 Empress Maria Theresia ordered the institution to be moved to Buda. Furthermore, the 'General Commission for the Supervision of Science' was created in 1795. It was the first advisory body solely created for the purpose of drafting science policy (Tamás 1985, p. 32).

The first half of the 19th century of European history is generally termed the 'Reform Period' or 'Vormärz'.[4] It is also the dawning of the (re)appearance of Hungary as a nation. During this time the feudal society was beginning to be transformed into a capitalist one (Berend/Ránki 1960). The steam mill was introduced to industry, rivers were regulated, the land under cultivation was expanded rapidly and roads were built (Heinrich 1986, p.7). The Jewish Emancipation Act finally allowed Jews to settle and buy property. In 1826 the Hungarian Association of Scientists was founded, later being renamed into the Hungarian Academy of Sciences (Magyar Tudományos Akadémia, MTA).

The MTA was, for the first 120 years of its existence, what Academies of Sciences are in the classical sense (U.S. National Academy of Sciences 1994):

> a private, non-profit, self-perpetuating society of distinguished scholars engaged in scientific ... research, dedicated to the furtherance of science and technology and their use for general welfare.

It was an organisation, whose members had been chosen for their reputation and which was free to elect new members amongst their peers. One of its main purposes during the 19th century was to foster Hungarian culture. It sponsored the translation of books into Hungarian, for the purpose of ending the domination of the country by German-speaking Austria. The organization was largely independent from the state until the 1930s. The institution's budget consisted predominantly of donations and estates (43%) and interest on funds (22%) in the first decades (1831–1880). From 1881 until 1930, however, its income pattern changed significantly to dividends from stocks and shares (24%), state aid (22%), income from rent (17%) and much less income from both donations and estates (16%) and as interest on funds (4%). As can be inferred from the figures, the state aid and with it the influence from government went up from 7% in the former to 22% in the latter period (Támas 1985, p.37).

After the bourgeois revolution of 1848, the peasants were granted freedom and feudal structures were weakened. Beginning with the 'Ausgleich'[5] of 1867, the Hungarian kingdom was granted more autonomy. The Habsburg Monarchy was reorganized into the Dual Austro-Hungarian

Empire. A market of more than 50 million people was all of a sudden freely accessible to Hungarian producers.

It is largely acknowledged by now that Hungary greatly profited from the 'Ausgleich' – perhaps more than the German speaking part of the empire. Count Gyula Andrássy, an influential personality of his times, in the following quotation writes that as a result of the better-developed political life and the comparatively more unified public opinion of Hungary until now (1897), his country came to have more weight than the Austrian position (Haslinger 1996, p.3):

> Ja, infolge unseres mehr entwickelten politischen Lebens, infolge der Einheitlichkeit unserer öffentlichen Meinung, ist unser Wert bisher mit größerem Gewicht in die Waagschale der Gemeinsamkeit gefallen als der Wert des politisch weniger einheitlichen Österreich.[6]

The inflow of foreign capital and technology, in combination with the huge market, led the Hungarian economy to the take-off stage.[7] Between 1867 and 1918 40% of the investments in the time span came from abroad,[8] mostly from Austria (Heinrich 1986, p.9).

A variety of authors have pointed out that the growth rates of the Hungarian economy were comparable to the growth of the classical and most successful late developer of the time, Germany (Pungor/Nyiri 1993, pp.25–39), or at least in line with the European average (Berend/Ránki 1985, p.18). One should keep in mind, however, that the industrial and technological bases of the industrialized part of Europe at that time were already on a considerably higher level than that of Hungary and that it is easier to achieve high growth rates from a low level than from a high one, relatively speaking. In addition, the Hungarian growth was not of the same quality as the growth of other European late developers: it was led by the lower value-adding agricultural sector and not by higher value-adding manufacturing sector. Exports consisted mainly of goods with a low technological content level such as agricultural raw materials and refined foods. For instance, Hungary supplied 23.7% of world exports in wheat flour – being surpassed in the production of this good only by the United States (Berend/Ránki 1985, p.22).

The technological development of the country was imperiled when industrialization, which was foreign led before WWI, slowed down after the break up of the Austro-Hungarian empire in 1918. The country shrank dramatically and so did the market freely accessible to the Hungarian producers. When the borders of Europe were redrawn after WWI, the number of people living in Hungary dropped from more than 20 million before the split up of the monarchy to less than 8 million after the break up and the subsequent proclamation of the Republic of Hungary. Universities, raw materials and whole industries were all of a sudden outside Hungary. The political as well as the economic system was able to stabilize itself from widespread polarization and hyperinflation only in the mid-1920s. Table 3.1 shows the rapid growth of student enrollment since the 1860s. It also shows the stagnation of student numbers in the 1920s and 1930s. Only after WWII can a dynamic development of student enrollment be observed again.

Table 3.1   Student Numbers in Hungarian Universities, 1850–1930

Year	Students
1850	838
1860	1.179
1870	2.629
1880	4.396
1890	5.218
1900	9.700
1910	12.951
1920	12.902
1930	12.611

Source: Tamás 1985, p.54.

On the one hand Hungary's development during the interwar era is often described as a time of stagnation. The educational system was not changed dramatically, in spite of the need to do so. As has been shown, the student numbers were not expanded, despite the backwardness of Hungary in this respect. The economic and technological base of the country progressed only slowly and industrial production barely reached the pre-WWI level. Moreover, the political system was volatile. On the other hand the dismal situational circumstances of these two decades should be taken into account – the loss of markets and people and therefore economic and political power.

Despite these circumstances one can see some progress during these two decades when taking a closer look. The minister of culture during the Bethlen governments, Kuno Klebersberg, and Zoltán Magyary were among the driving forces behind reforms. In 1927 Magyary edited the volume 'A Magyar Tudománypolitika Alapvetése' (The Foundations of the Hungarian Science Policy) describing foundations and future directions of the Hungarian science policy.

Universities were founded and others enlarged. Advances of the S&T system included the first industrial laboratories, which were set up primarily in the growth industries of electro-technology, pharmaceuticals and certain engineering units (Tamás 1985, pp.38). Research contacts between universities and industry were slowly established. However, only one chair, for nuclear physics, was financed by a Hungarian firm, Egyesült Izzó, later named Tungsram.[9] Despite its slow pace, this development certainly meant progress for Hungary. It was, however, counterbalanced by a brain drain that first affected leftists and Jews, then, directly after WWII, right-wing Hungarians, followed by many other non-communist intellectuals between 1945 and 1948 and, finally, in 1956, conservatives as well as reform communists.

The reason for the first of these waves of people leaving the country was the coming into power of the fascist leader Gömbös in 1932. During the 1930s coalitions first with Austria and Italy, later Germany and Italy dominated the economic and political orientation of Hungary. In terms of its political structures the country was comparable to the other Middle European fascist states. To an extent this could be said already for the 1920s, because corporatist structures had, though often not elaborate in nature, preceded fascism (Janos 1982, pp.222).

Technologically, the country was increasingly dependent on Germany, as is true for others in the region, especially Austria. Particularly the growth in the technology-intensive heavy industries was built upon a substantial foreign debt, which again was drawn primarily from Germany.[10] An indicator for the dependence of the small country Hungary on the large power Germany was that in the second half of the 1930s, 50% of Hungary's trade was with Nazi-Germany (Heinrich 1986, p.23).

During the war Hungary displayed strong features of a command economy, favoring certain industries and production lines and fostering R&D in sectors that could be directly utilized, as for example in telecommunications. If one puts aside the incredible losses of lives and the repugnant and barbaric nature of warfare, one can conclude that technologically Hungary profited in a number of branches from the War effort. Bauxite production was introduced to the country on a large scale. The existing aircraft industry was greatly enlarged. In 1944, for instance, 537 Messerschmidt planes were produced in Hungary (Berend/Ránki 1960, p.130). The value of total production of manufacturing, as applied to the Hungarian territory of 1938, rose in constant prices by 37.5% from 1938 to 1943, with a 20.7% increase in 1939. The numbers of employees in manufacturing, again applied to the territory of 1938, rose by 35.7% (Berend/Ránki 1960, p.140).

However, WWII caused losses, not only from a human(istic) standpoint, but also from an economic and technological point of view. For example, there was a severe loss of professionals – a third of all the medical doctors of the country were lost by 1945 (Tamás 1985, p.39). And although the productive capacity of the Hungarian industry in 1945 was higher than in 1938, the damage done to the infrastructure was severe and has to be weighed against the production gains.

3.2 Policy Maker's Models and Intentions

The period from the beginning of the intensive phase of industrialization in 1867 until the end of WWII can be divided into two phases, each ending with a World War. First, the industrialization phase from 1867–1914, which extended into WWI until so much blood was drained from the economy that the industrialization came to a standstill and was even turned back to some extent. As has been established, the phase of 1919–1938 has, with respect to the further industrialization of the country, brought first a recession, then a stabilization on a lower level, followed by a mild upswing. During WWII we find first a surge of industrial development, and then, paralleling the situation during WWI, a stand-still of industrial development. Because of the long-term effects on the S&T system, the by far more interesting period for this study is the phase of intensive industrialization from the 'Ausgleich' until the outbreak of WWI. Consequently, the time span from 1867 till 1914 will be analyzed during the next pages.

Hungarian policy makers during the second half of the 19th century primarily strove for an enlargement of the country's autonomy. This effort was paralleled by an understanding that economic development was a necessary prerequisite for such a process. While there are no clear signs that a linkage between scientific and economic progress was comprehended by policy makers, the fostering of technology indicates an understanding of the linkage between technological change and economic development.[11]

A number of strategies were used by the Hungarian governments to foster economic growth in direct and indirect ways. During the last three decades of the 19th century at first light, but then increasingly heavy[12] industries were granted privileges in Hungary. Since the sheltering of sectors through tariff barriers was not allowed by Austria, the Hungarian government tried to induce growth primarily through subsidies, tax allowances and procurement policies.

In fact, Austrian business people were soon envying their Hungarian peers for the active role of the Hungarian state. Since 1867 the Hungarian economy was growing fast, with the state fostering and partially also steering this growth. The rapidly expanding Hungarian train-system, for example, was almost fully state-owned by 1891 – a situation quite different from Austria, where important parts of the train-system still were private – and undercapitalized. But not only infrastructural measures were taken to instigate economic growth in the country: both, technology transfer from foreign countries and the internal transfer of the Hungarian economy were aided in various ways, too. The Ministry of Trade, for instance, disposed over a fund for the acquisition of engines and tool making machines for small and medium enterprises.[13]

The enacting force from the central government for the technology system was the Ministry of Trade. It disbursed funds in order to foster industrial development, partially without being regulated by law in doing so. Table 3.2 provides an overview of those parts of the ministry's subventions around the turn of the century, which were not regulated by law.

Table 3.2   Subventions to Industry from the Ministry of Trade, in Million Kronen, 1868–1914

Time Span	Yearly Average
1868 – 1880	0.03
1881 – 1890	0.13
1891 – 1900	0.56
1901 – 1909	4.42
1907 – 1914	7.06

Source: Figures cited from Paulinyi, Akos 1995, p.202.

Note: The total sums spent on industrial development during the time span were significantly larger – these figures reflect the dispersion of public funds through the Trade Ministry without any regulation through laws.

Although the S&T system in general was weakly regulated, a number of activities in the sphere of economic and technology policy of the Hungarian state were regulated through laws. A starting point was law number 44 (Gesetzes Artikel Nummer 44 aus 1881) passed by parliament in 1881. In this law a number of production branches, which featured 'appropriate technologies' were to be aided through special regulations. The passage addressing the 'appropriate technologies' was also included in the laws subsequently issued to regulate industry. Specifically, the textile, metal and machinery industries were exempted from taxes under law 44. The decision on the question of which company actually featured 'appropriate technologies' were up to the Ministry of Trade and the Ministry of Finance (Paulinyi, 1995, pp.180).

During the next years the activities of the Ministry of Trade grew substantially. First, capital was pumped into companies which were in trouble but judged as economically and strategically valuable. Then, for a number of companies fares on the public railway system were reduced for the transport of construction materials and machinery (Paulinyi 1995, p.181). In addition, land was given to companies seen as promising (Haslinger 1996, p.85). Moreover, funds for the fostering of economic growth were enlarged. Finally, procurement policies were used for increasing demand in strategic areas as for example in the defense industry.

As a result of the armament race preceding WWI, Hungary's technology system was experiencing a serious push into industrialization. Warship production was begun with credits under the Monarchy's warship building program in 1911. Of the program's million 312 Kronen Hungary lucrated million 113 Kronen. Moreover, the first aircraft industries, set up in 1912, were solely producing for the military. The same can be said of manufactures such as the Manfred Weiss works, which had played an important role in the beginnings of the industrialization of the country. In 1913 this manufacturing company produced military equipment in large numbers with 5000 workers and machinery representing 20.000 horsepower (Berend/Ránki 1960, pp.18) – more than double the amount of the total power supply in Hungarian industry 50 years before (Berend/Ránki 1960, p.2).

The state's role in the industrialization process is also indicated by the industrialization bill of 1907, a successor of Article 44 of 1881. As a result of the law million 38 Kronen were dispersed as subventions to industry in the time span from 1907 to 1914. 460 manufacturing companies were granted tax allowances, 121 factories were newly built utilizing governmental funds, 190 factories were enlarged and 380 received new machinery (Berend/Ránki 1960, p.12).

The Hungarian policy-makers' conceptions of economic and social development in the decades after the 'Ausgleich' were strongly influenced by the models of the early industrializer England and the second wave of industrializing countries with Germany as the most successful example. This becomes especially visible in the reform attempts of the Hungarian S&T system. German institutions featuring technical functions were flourishing during industrialization – this is especially true for the higher education sector.[14] Paralleling this development, in 1871 a technical university was founded in Budapest in order to provide the technical expertise dearly needed in the country.[15] In 1872 another university was created in Klausenburg and before the beginning of WWI a third university was set up in Kolozsvár, today's Romania. In addition, existing higher education institutions were substantially enlarged. A number of disciplines were introduced to the Hungarian S&T system, others were fully established.

During the last three decades of the 19th century non-university research institutions were also founded. Most of them were grouped around the agricultural industry, as was the case with the Seed-Grain Research

Institute and the Agricultural Machinery Experimental Institute. Another example is the National Phylloxera Experimental Institute, which was founded as a reaction to the vine-pest in 1874 (MTA 1985, p.6). Others were the National Meteorological Service, created in 1870 and the Institute of Psychology, established in 1902 (Farkas 1985, p.91). However, these institutions were not concentrating on original research. Among their day-to-day operations was the control of food quality and similar services (Palló 1995, p.277).

Not only were universities and institutions founded, but a decisive increase in students could be observed, too. While in 1870/71 6.363 students were enrolled in Hungarian institutions of higher education, in 1903/4 the number of students had more than doubled to 12.822.[16] In addition, Hungarians continued to visit foreign educational institutions, predominantly in other countries of the empire. Growth rates in the humanities were distinctively higher than in the natural sciences.

The ideal-typical university as introduced by the German Alexander v. Humboldt earlier in the century was the ultimate goal of the Hungarian policy-makers' and scientists' endeavours. The high esteem of this type of universities, unifying teaching and research functions, is reflected in the following statement of the minister for culture Gyula Wlasics. On January 24, 1895, the minister first emphasized his low opinion of the French and English university system and then said that only the German university system raises the university to its high task of teaching and engaging into research:

> Das deutsche Universitätssystem ist es, das die Universität auf das Niveau ihrer hohen Aufgabe hebt, da die Universität nicht nur eine Lehranstalt, sondern auch eine Wissenschaft betreibende Institution ist. (Vámos 1995, p.218)

While it is not clear how many institutional solutions in the S&T system were imported from Austria and how many from Germany,[17] it can be safely stated that the influence of German technology, in the sense of technical as well as organizational knowledge, on Hungary was large during the time of the industrial revolution. One example which demonstrates this is the strategically important sector of tool-making machines. Since tool-making machines are used for constructing machinery in all kinds of branches, they are as important as they are technology-intensive. The Hungarian imports in tool-making machinery during industrialization were growing substantially. Between 1893–1897 and 1909–1913 the imports grew by a staggering 832%. Since Austrian firms neither had the technology nor the capacity for the satisfaction of such growth, the German share of Hungarian imports of tool-making machinery rose during this time from 26% to 47% (Paulinyi 1995, p.191). In the light of the Austrian tariffs sheltering the Austro-Hungarian Monarchy from world markets, this growth was all the more impressive.

3.3 Actual Effects of Policies

The result of the evaluation of Hungarian economic and technology policies during the industrialization of the country is twofold. On the one hand, one has to acknowledge the interdependence between autonomy and prosperity in Hungary during that time. The increased autonomy Hungary enjoyed was as much an effect of the economic strengthening of the country as the economic prosperity was a result of the autonomy.[18] In this respect Hungary had made good use of the historical situation, with policy makers aiding this process. After all, the country's GDP rose between 1867 and 1913 from million 1330 to million 5064 Kronen, an average growth of nearly 8.3% per year (Paulinyi 1995, p.176). On the other hand, it has to be pointed out that despite the high economic growth rates in the years after the 'Ausgleich', Hungary had not succeeded in being industrialized on the eve of WWI, with 61% of the GDP still being produced by the agricultural sector, only 28% by industry and mining and 11% by commerce and transport (Paulinyi 1995, p.174).

A fairly good measure for the degree of industrialization is the number of employees working in industry. Table 3.3 provides evidence for the speed of industrialization around the turn of the century.

Table 3.3   Number of Employees in Industry and Mining, 1890, 1900, 1910

Year	1890	1900	1910
Employees	156.581	281.416	464.491

Source: Paulinyi 1995, p.201.

Note: Ventures with more than 20 employees are included.

These impressive growth figures are mitigated by several factors. First of all, the starting level in 1890 was low with barely 157.000 employees in industry and mining in comparison to the 20 million inhabitants of Hungary at the time. Then, when analyzed according to sectors, the distribution of growth occurred unevenly, as is shown in table 3.4. The data, which are most disturbing for an industrializing country, show that textiles and chemical products grew substantially around the turn of the century, but the most essential production of machinery was shrinking.

Table 3.4   Shares of Branches in the Total Output of Manufacturing and Mining Industries, 1898, 1913

Branch of Industry	1898	1913
Mining	6.7	5.4
Iron and Metals	12.6	15.2
Machinery and Electrical Goods	11.4	9.2
Chemical Products	5.7	7.3
Building Material	3.5	4.1
Textile and Clothing	5.0	7.3
Leather	2.1	2.7
Timber	6.6	6.8
Paper and Printing	2.3	3.1
Food	44.1	38.9
Total	100.0	100.0

Source: Berend/Ránki 1960, p.27.

As can be inferred from the table above, the production of metals and iron was rising around the turn of the century. However, the growth in this field was neither comparable to German or Austrian growth rates nor was it capable of satisfying the needs of the home-market. While pig-iron production in Hungary rose by 32% between 1899 and 1913, production numbers in Austria rose by 76% and in Germany even by 164%. Between 1897 and 1900 production covered nearly 93% of consumption in Hungary, but in 1913 only around 50% of the demand (Berend/Ránki 1960, p.17).

Moreover, despite the growth in parts of Hungary's heavy industries, light industry, especially in the form of foodstuff production, was still dominating in manufacturing. Table 3.5 shows how uneven the distribution of industrial production was in Hungary shortly before WWI.

Table 3.5 Distribution of Branches Within Manufacturing of Early and Late Developers, 1913

Branches of Industry	England	Germany	Austria	Hungary
I. Iron, Metal, Machine and Chemical	32	45	28	28
II. Textile, Leather and Clothing	26	19	29	11
III. Food	13	17	27	46
Resulting Ratio of Light to Heavy Industry	1.2:1	0.8:1	2:1	2:1

Source: Berend/Ránki 1960, p.28.

As can be inferred from table 3.5, the German and English economies before WWI were already fully dominated by heavy industries. In the case of Austria and Hungary light industries still were exceeding production in manufacturing. With respect to all other countries in the table, Hungary featured a predominance of foodstuff production, which is the technologically least advanced and least value adding of the three groups of branches reflected in the table.

Hungary's heavy industries were, in comparison to the other branches, newly established and technologically advanced. However, the country's heavy industries were small in comparison to Austria, as was the whole manufacturing sector. The output per capita of Austria's heavy industries before WWI was 75% higher than that of Hungary (Berend/Ránki 1960, p.29). When analyzing the Monarchy's economy as a whole, it is interesting that shortly before WWI, with around 40% of the Monarchy's population, Hungary was covering only about 25% of the Austro-Hungarian industrial production (Berend/Ránki 1960, p.32).

The development of the science part of the Hungarian S&T system was similarly plagued with bottlenecks. At the end of the century the country of almost 20 million people had three real universities,[19] of which, as mentioned before, two were newly built: the Technical University in Budapest and Kolozsvár University, in today's Romania. The focal point of the Hungarian higher education institutions was Budapest, which in 1896 featured universities with a total enrollment of 4002 students, only slightly less than could be found in Vienna, which had 4536 students at that time (Vámos 1995, p.220). However, in the case of Hungary this meant not only many, but 'too' many students for the existing infrastructure. We find reports of a lack of assistants (assistant professors), lack of space and an aged infrastructure (Vámos 1995, pp.217).

This situation and a state increasingly ready to invest money into the knowledge of promising students in the form of government grants led to a growing number of Hungarians studying abroad. In this respect Hungary's orientation towards Germany is highly interesting. From 1881 to 1914 more than 8000 Hungarians studied in Germany with stipends from their government, of which almost 2800 visited Technische Hochschulen (today's technical universities). During the same time span the percentage of Hungarian students visiting Viennese universities sank from 9.4% of the total number of students in Vienna to 1.6% in 1910. At the university of Graz the share of Hungarian students of the total number of students at the site went down from 3.2% in 1863 to 0.3% in 1910 (Vámos 1995, pp.221). Similar to the situation in key technologies, the importance of German science for Hungary rose during the last decades of the Austro-Hungarian Monarchy to the same extent as the influence of the German-speaking of the Monarchy waned.

3.4 Possible Explanations for the Differences

The results of this inquiry into the nature of Hungarian economic development from 1867 to 1914 are: Policy-makers planned a phase of development, which should have led the country into industrialization and autonomy from Austria. Indeed, there was an economic spurt, but it lead not to full industrialization, with the concomitant evolution of S&T, but rather to a mixed agrarian-industrial economy, with a rather mediocre S&T system. Are the reasons for this development connected with S&T? And: if yes, which reasons connected with S&T are discernible?

A number of possible causes for the only half-accomplished industrialization of Hungary are closely connected with questions of S&T. A prime reason might be found in the autarchy of the Austro-Hungarian

Monarchy. While the tariff barriers inside the Monarchy were rather low, the economy was quite closed off from foreign countries. Hence, the permeability of Hungary's borders for technologically advanced machinery was not sufficient enough to still the demands of the expanding economy. Hungarian policy-makers from an early stage on realized the dangers and costs of this autarchy for the development of their country. This problem was also part of the reoccurring disputes over the economic relations between Austria and Hungary.[20] After all, the costs burdened on the Hungarian buyers of machinery were high. Although Austrian machinery in comparison to equivalent products from foreign countries often was more expensive, after customs the foreign products were mostly costlier than the Austrian machines. This, of course, hampered the industrial development in Hungary, but helped Austrian machinery manufacturers to gain profits (Paulinyi 1995, p.191).

Looking at the most important Central European late-industrializer of the time, Germany, one can see that at least two ingredients essential for industrialization were generally weak inside the Monarchy, but almost completely missing in Hungary. The first lacking factor is capital, the second technology.[21]

The lack of capital and technology inside Hungary led to a strong foreign involvement in the industrialization of the country. As has been mentioned before, in the time span between 1867 and 1918 forty percent of the investments were coming from abroad,[22] the largest part from Austria (Heinrich 1986, p.9). Over time, the ratio of foreign to indigenous capital changed in favor of Hungarian funds.

Table 3.6   Share of Foreign Investments in Hungary from 1867–1913

Time Span	Foreign Capital
1867 – 1873	60%
1873 – 1900	45%
1900 – 1913	25%

Source: Vámos 1995, p.261.

It is questionable whether foreign investors acted in the same way – for instance with respect to the time horizon of their investments or the rate of profit-reinvestment – as Hungarian investors would have done in the time of industrialization. A further inquiry into this matter would be specifically interesting in the case of the large banks, which were strongly influenced by foreign capital (Berend/Ránki 1960, p.37). After all, banks played – and in a number of these countries still play – an important role for the development of the late industrializers, especially Germany.[23]

For the second factor scarcely represented in Hungary at the time of the 'Ausgleich', technology, imports played a strong role, too. It has been mentioned that the import of machinery was extensive during all phases of industrialization between 1867 and 1914 (see Paulinyi 1995). This fact in itself would not have been hurtful to the Hungarian economic development. However, it can be seen as problematic that in many branches Hungarian machinery production could not gain market shares over time. In 1909 the production of machinery for the leather, paper and sugar industry covered a minuscule 1–2% of demand. In 1912 a meager 20–30% of the comparatively large textile industry's and 30% of the metalworking industry's demand for machinery could be met by Hungarian production (Berend/Ránki 1960, p.19). As the production of capital goods (goods, other than raw materials, usable for producing other goods) could not catch up with the production of consumer goods (goods, which have as ultimate function to be consumed), it can be concluded that the Hungarian economy was not capable of technologically catching up with its Western neighbors.

Moreover, the technological level of Hungarian personnel was quite low in comparison to the late-industrializers or even Austria. Indications for this statement can be found in the fact that a large number of foreigners were working for large companies[24] in Hungary around the turn of the century, as is shown in table 3.7.

Table 3.7   Share of Foreigners Working in Large Companies in Hungary, 1900, 1910

Functions	1900	1910
Owners	4.4	3.4
White Collar Workers (Trade, Bureaucracy)	10.0	7.1
Technicians	20.7	16.6
Foremans, Mechanicians	19.3	12.5
Shop Floor Workers	10.2	5.7

Source: Vámos 1995, p.262.

As can be inferred from table 3.7, despite the falling levels of foreigners working in Hungarian industry, the share of foreigners was still comparatively large in the technologically most demanding job category – the technicians. But it was not only the low level of technological knowledge which was hurting the industrialization of Hungary. Another problem was the fact that the Hungarian technical intelligentsia was often not employed by Hungarian firms.[25]

Still another factor hampering the economic and technological advancement of Hungary was the structure of society, which, especially in the 19th century, still bore many remnants of the feudal system. The strength of the agrarian sphere of society strangled the efforts of the new bourgeoisie to modernize society after a capitalist fashion. A striking example for the distortion of the Hungarian industrial policy by the agrarian faction is the usage of a sizable part of the aforementioned subventions to industry by the Trade Ministry: these monies were used to build distilleries on the big estates of the large landowners.[26]

One more factor hindering Hungary's catching up with the industrialized countries was the blossoming of cartels and monopolies, beginning with the iron cartel in 1879. The oligopolistic cartels as well as monopolistic companies controlled markets and prices and fixed production rates on the basis of the output of the last years. Regularly, foreign companies, led by Austrian firms, were involved in these organizations (Berend/Ránki 1960, pp.33). Competition, generally recognized as a driving force of technological advance, must have been hampered by cartels and monopolies.

In both, science and technology, Hungary was lagging behind Austria at the time of the 'Ausgleich' as well as on the eve of WWI. However, the development of the science and technology subsystems in each part of the Monarchy show different trends for these 47 years. It is not entirely clear if the distance between the two countries was really getting any smaller on the level of technological capabilities. On the one hand, Hungary was making large steps into the direction of industrialization, on the other hand the growth rates in strategically important branches were not dramatically higher[27] in Hungary than in Austria. Similarly, one can observe that the educational level of Hungarian technical staff was increasing vastly during the last decades of the Monarchy. Nevertheless, the influx of technical personnel from foreign countries was steady.

In the science part of the S&T system, however, there are at least two factors which can be seen as a success for the Hungarian strive for scientific advancement in the late 19th and early 20th century. Firstly, the afore-mentioned rise in numbers of students and schools are proof for such a success. Of crucial importance for the scientific development of the country was the sending out of students into centers of excellency in Germany and Austria, who returned with knowledge on the newest research, thereby strengthening the knowledge base of the country markedly.

Secondly, Hungary during the time from 1867 to 1914 was gaining ground insofar as it was laying the bases for the creation of a scientific intelligentsia, which, during the 1920s and 1930s was at least as successful as Austria's science community.[28] Scientists as Zoltán Bay, Theodor Kármán, John Neumann, Michael Polányi, Albert Szent-Györgyi, Leo Szilárd or Edward Teller were educated in Hungary during this time. A generation of exceptional scientists was working in Hungary, before many of them, of their own free will or with some help from the authorities, left the country – among them eight Nobel laureates in physics, chemistry and medicine.

Notes

1   Alexander Gerschenkron's classical analysis of the industrialization of Europe during the 19th century (1962) distinguishes nations industrializing during the beginning of the century, termed 'early developers', as for example England, and nations industrializing during the second half of the century, 'late

developers', as Germany. Gerschenkron ascertaines that the Austro-Hungarian empire did not get industrialized during the 19th century (1977), it therefore could be termed a 'late late developer'.

2   In rankings done by the MTA's Information Science and Scientometrics Research Unit (ISSRU) Hungary persistently is among the top 30 countries of the world with respect to publications output and citation impact of a variety of S&T disciplines (Braun 1994, pp.299).

3   In Hungary, since the 1980s, there has been a renewed interest in the national history of science. As one interview partner pointed out, this subject was not opportune during realsocialism and its stress of the virtues of internationalism. For accounts of the history of the nation's S&T, (see for example Füzeséri 1992 and Szabadváry/Vámos 1994).

4   A German expression, literally 'before March'. In March 1848 all over Europe bourgeois revolutions broke out. The Hungarian revolution could be suppressed brutally only in 1849, after Russia sent a corps against Hungary, thereby assisting the Austrian Habsburgs dynasty in fighting the revolutionaries.

5   German for 'compromise', Kaiser Franz Josef II was forced to pursue this strategy for political and economic reasons having withstood the pressure to do so for several years. The main reasons for the 'Ausgleich' were the lost wars against Italy 1859 and Prussia in 1866, which forced Franz Josef to give in to the reform efforts and to grant, amongst other things, the first Austrian constitution.

6   Austria was more split than Hungary due to the multi-ethnic composition of this part of the empire as well as the increasing democratization, especially of the German speaking part of the country.

7   The term 'take off' is part of Walter Rostow's theory of economic development, formulated in 1961, characterized by an industrial revolution generating steady growth, thereby leading an economy from an underdeveloped to a developed industrial level.

8   Exceeding Heinrich's estimation (Heinrich 1986, p.9) by 10%, Berend/Ránki insist that even more, 50%, of all the capital invested into Hungary in this time span came from abroad (directly and indirectly, i.e. through banks, investment etc.); see: Berend/Ránki 1985, p.18.

9   Inzelt, Annamária (ed), 'Institutional Support for Technological Improvement – Back to Cooperation from Rigid Seperation', draft version of a World Bank project, typescript, p.31.

10 For an analysis of German trade policies in Eastern Europe from 1890 to 1990 and the politics behind the policies, see Spaulding 1991, pp.343; for an analysis of German economic policies driving Austria towards the 'Anschluß' beginning with 1918, see Biegelbauer, Peter, 'Foreign Participation in

Austria's Economy', typescript prepared for the international project 'Europe in the Time of the World Economic Crisis', University of Vienna, 1990.

11   For a Hungarian policy-maker of the 19th century the connection between science, technology and economic development must have been less straigthforward than it seems today. After all, the first industrial laboratory in Hungary was created in the early 20th century and even in England, Germany and the United States industrial R&D was institutionalized only from the 1870s on.

12   See also Gerschenkron's classic chapter 'Economic Backwardness in Historical Perspective', written in 1962, as well as his book on the subject in 1977. In the latter book Gerschenkron cites a letter of the Austrian Marxist Victor Adler to Friedrich Engels, in which Adler happily states that 'Hungarian industry is raised by the state' (Gerschenkron 1977, p.28). Adler hoped that the growth of an indigenous proletariat in the empire soon would lead to a Marxist revolution.

13   Haslinger quotes this information originating from the petition of the Lower Austrian Chamber of Commerce to the Austrian Minister of Trade (Haslinger 1996, p. 83).

14   Otto Keck states that while in 1891 in Prussia less than 2000 students were enrolled in Technische Hochschulen (the equivalent to todays's technical universities), in 1911 there were already 4064 students. At the same time student numbers in mid-level education were nearly quadrupling from 1016 to 3760. See Keck 1993, p.122.

15   Eva Vámos gives this date, whereas András Füzeséri gives 1870 as the starting point for the new technical university. See Vámos 1995, p.217 and Füzeséri 1994, p.7.

16   Elekes, D., 'Budapest szepere Magyarország szellemi életében' (Budapest's role in Hungary's intellectual life), Budapest, 1938, p.45, cited in: Tamás 1985, p.35.

17   It is also unclear how many innovations were taken to Austria from Germany and then taken to Hungary from Austria.

18   This observation is reminiscent of the results of the small states research of the 1980s, when the analysis of the economic success of the small open Western European economies was the object of a number of books. See, for example, Katzenstein 1984 and 1985.

19   Most of the colleges were theological seminars.

20   For a detailed and exemplary description of one of these disputes in 1895–7, see Haslinger 1996, pp.80.

21   Land and labor, the other variables of a typical production function, were abundant in Hungary at this time.

22   Berend/Ránki speak of a rate of 50% – thereby advancing an estimate 10% higher than Heinrich's (1986) figures; see: Berend/Ránki 1985, p.18.

23   On the role of the banks, see especially Gerschenkron 1962. For a more recent appraisal of different economic governance regimes or capitalist models, with an emphasis on the central role of capital markets and banks, see Albert 1993.

24   Large companies are of special interest for the analysis of an economy's technological capabilities as, on an average, they are technologically more advanced than small and medium sized companies.

25   Akos Paulinyi mentions this fact (Paulinyi 1995, p.184). A reason might be that technologically sophisticated firms were dominated by Austrian and German engineers keeping to themselves and machinery often brought or ordered by these engineers again demanded Austrian or German engineers (see Vámos 1995, p.260).

26   Berend/Ránki even write that the 'majority' of the funds allocated after Article 44 of 1881 were used for the creation of distilleries (Berend/Ránki 1960, pp.11).

27   For a catching up with Austria, Hungary, starting from a quite low level of economic development in 1867, would have needed strong and persistent growth featuring a markedly higher expansion than in Austria.

28   One could quite easily make the argument that Hungarian scientists during the interwar period were even more successful than Austrians.

4 The First Transition: The Implementation of the Soviet S&T System

4.1 Historical Development of the S&T System

Shortly after WWII the Hungarian S&T system still was organized along the lines of the German model, which had been imported directly from Germany and, during the last decades of the Austro-Hungarian Monarchy, via Austria. The small S&T system[1] included universities which combined the functions of teaching and research as well as an industrial R&D system organized and sponsored by private sources. As a result of the dearth of capital after WWII, Government played a diminished role, utilizing instruments such as procurement and subsidies.

After the end of the War and the election of 1947, the Communist Party gradually enlarged its influence with the help of the Soviet Union. The proclamation of a new constitution by the Communist government in 1949 was a stepping stone towards its total control of society, as was the construction of detention camps. The direct results of this development for the S&T system was a renewed brain drain as well as a transformation and centralization of most institutions.

Between 1949 and Stalin's death in 1953 the General Secretary of the Communist Party, Mátyás Rákosi, was the totalitarian leader of the government. The changes he implemented led to the establishment of a direct copy of the Stalinist Soviet model, with its extreme emphasis on the heavy industries and the centralization of powers in the Party. The model at this time looked like a successful road to industrialization – given both the ability of the Soviet Union to withstand the German armies and the rapid build up of capacities in heavy industries during the 1930s and 1940s. Rákosi's political adversaries were imprisoned. The profits of agriculture were used to build heavy industry. The S&T system was, as is true for many subsystems of society, compartmentalized, centralized and planned. As has already been mentioned in the introduction, science is perceived by Marxists as one of the driving forces for the restructuring of society. In fact, it will be shown that in the realsocialist People's Republic of Hungary science was valued so highly that budget constraints were of negligible influence to the S&T planners until the early 1970s.

In 1948, the government created the Hungarian Council of Science, in the language of realsocialism, the 'supreme body' of science. The institution's role was to reorganize the structures and functions of the science system. The body was headed by Ernö Gerö, the number two (after Rákosi) of the Communist Party. At this time, the possibilities of resistance within the S&T system were already minimized. Apart from the continually tightening political control, the Academy of Sciences (MTA) was also deprived of an important income source, namely its landed estates. In addition to this, the scientific community of the country was deeply split into one group including the humanities and social sciences and another group, consisting of the natural and technical sciences. As a result of this cleavage in the S&T system the Communist Party could employ a 'divide et impera' strategy, thereby weakening the resistance of the scientists.[2]

It is interesting to find that the MTA was one of the main beneficiaries of the new emphasis on science. The original plan of the Communist leadership had been to let the MTA die a slow death because of its conservative nature and to centralize all science policy functions in the newly founded Council of Science (Péteri 1993, pp.291). This plan, however, was not to become reality.[3] In the early 1950s the MTA was remodeled after the Soviet S&T system which had its organizational power centralized in the Academy of Science. Soon the MTA grew at an astonishing speed. A whole series of institutes were set up, with the budget and the number of personnel expanding. Most of the growth occurred in the natural and technical sciences. The Academy founded new institutes, like the Institute for Nuclear Research in 1954, and took over existing ones, like the Biological Research Institute in 1951. The humanities and social sciences were only supported reluctantly, either because of the role they played in resisting the Communists or because they only had an indirect effect on the economy, only supported reluctantly. During Rákosi's era only one institute in the fields of humanities and social sciences was founded: the Research Group for Folk Music in 1953.[4]

After the death of Stalin, Rákosi's gradual loss of power and the loosening of the state's grip on society led to the gradual opening of the country. During this time the Hungarian scientists were subject to less political control than during the first years of the Rákosi regime. Despite the suppression of the revolution by Soviet troops in 1956, the climate for research became more open with the end of Stalinism, at the same time political control gradually loosened.

The MTA, as holds true for the whole S&T system, was less disturbed by the events of 1956 than the rest of society (Tamás 1985, p.43). Still, the political crisis that killed 25.000, injured 150.000 and led to the immediate emigration of 200.000 people, did not leave the Academy unharmed – especially since intellectuals were very much involved in the uprising. A number of researchers were expelled from the MTA,[5] the newly found freedom to reopen discussions in fields such as neurology, systems theory and management theory was lost again. The general political atmosphere, however, never became as oppressive as it had been in the years of the Rákosi regime again. After 1956, the new General Secretary of the Communist Party, Janos Kádár, sought a course of reconciliation and slowly displaced the Stalinist leaders. The pragmatic course of the Kádár regimes finally led to the economic reform initiatives of the 1960s, which will be the subject of the next chapter.

4.2 Policy Maker's Models and Intentions

For an analysis of the models behind the reorganization of the Hungarian S&T system in the aftermath of WWII, one has to take a closer look at two developments: the reaction of the intellectuals in Eastern Europe towards the failed attempts to catch up with Western Europe economically and the events in the Soviet Union since its establishment in 1917.

In a situation in which there was a series of crisis, one following upon the other, communism provided a seemingly successful ideology propagating modernization through intensive usage of the factors of production, steered from a central planning unit. During a large part of the 20th century this ideology was especially tempting for intellectuals in the backward economies of the European semi-periphery. After the failure to achieve an economic spurt under a capitalist system, the time seemed to be ripe for another drive for modernization, this time under the constraints of a new system: socialism. Already in June 1919 Béla Kun, leader of the

Hungarian Communist Party, raised these issues during the short-lived Soviet Republic:

> The slender chances provided by the collapse of the capitalist mode of production for the development of the forces of production can be augmented by force, by the new force of the proletarian state .... (The) lower level of economic development can certainly be compensated for by the use of force, the force of the proletarian state, ... By using the repressive organization of the state we can now achieve in a year or two what could not be achieved in twenty-five years of capitalist development, because of capitalism's immaturity (in Hungary). We can concentrate production to a degree that exceeds even the concentration by American capital amalgamations. In what I hope will be a short period of dictatorship we can attain artificially, through the force of the proletarian state, what we have failed to achieve in fifty years of capitalist development. (Péteri 1984, p.113)

Twelve years later, in 1931, Josef Stalin, in his speech to the first federal conference of the functionaries of socialist industry, said along similar lines,

> (w)e are behind the developed countries by 50–100 years. We have to eliminate this gap in ten years' time. Either we succeed in eliminating the gap or they will trample down us [sic! PB].[6]

As has been emphasized, this notion of the importance of catching up with Western Europe is central to an understanding of the development of Hungary. Indeed, the idea of utilizing modernization in order to eliminate the national economy's backwardness by means of the socialist organization was alive until the last years of the People's Republic. This opinion of a number of interview-partners is reflected in a booklet of the MTA on the national R&D effort in the mid-1980s, which states dryly:

> Essentially, it is since 1949 that Hungarian science has taken a socialist direction, implying a new type of institutional relationship between science and economy. This development was justified not only by the relative backwardness and the need to make up for the war losses, but also by the modern requirements of the scientific and technological revolution taking shape throughout the world. (MTA 1985, p.8)

This process had been started in the Soviet Union much earlier, with the take-over of power by the Communist Party. In the early days of the

Soviet Union the need to modernize and to strengthen the economy was a matter of survival for the young state, which was internationally isolated and ravaged by WWI and the subsequent war against the White Guards. Science and technology were seen as vehicles for the modernization project of Socialism. Under the constraints of the socialist notions of planning, the S&T system of the Soviet Union had to be rearranged completely. During the 1920s a lively discussion was waged amongst the party elite of the Soviet Union about the way R&D should be organized.

A number of planners were sent out to observe the S&T systems of the leading industrialized nations, England, France, Germany and the United States. Most models were more or less completely disregarded. The organization of R&D in the United States was envied for the strength of its industrial research, but at the same time disregarded as wasteful due to the competition of a number of companies in most fields (Graham 1993, p.174). On the contrary, in the Soviet Union R&D should be carried out in large centralized institutions making their results available to everybody concerned with the application of the new knowledge. Clearly, the underlying notion here is 'pipeline' notion of innovation, the basis for the 'science push' policy paradigm.

Finally, Germany was chosen as model to build upon. The 'Kaiser Wilhelm Gesellschaft', with its research institutes, which were independent from the university system and its orientation on basic research with some industry cooperation, was seen as promising. However, the decentralized nature of the German institutes stood in apposition to the very idea of a centrally planned economy. A solution was found in the centralized structure of the Prussian Academy of Sciences. Of course, the restriction of the Prussian Academy of Sciences within its old functions as a learned society, which was not directly immersed in scientific research, did not really befit the demands of the Soviet Union. Therefore, the Academy of Sciences in the Soviet Union, an enlarged and propped up version of the Prussian Academy, was to be the umbrella organization for the research institutes modeled after the Kaiser Wilhelm Gesellschaft.

Under these circumstances the role of the universities had to be changed, too. Instead of using the classical model of the research and teaching universities, which were dominant in Weimar Germany, the higher education institutions were restricted to education only. With this move, a problem that had plagued the science planners could be solved easily: on the one hand, the scientific elite of Russia was considered as bourgeois by the Communist Party. On the other hand, it was clear that the Soviet Union could not create the S&T base it longed for without scientists from the Soviet Union. Now staff judged as politically not trustworthy, but capable of doing good research, was put into the Academy.[7]

The S&T system of the Soviet Union finally was in place in the mid 1930s. It consisted of three building blocks. One block was the system of the Academies of Sciences, including the Academy of the Soviet Union, a net of regional and several specialized Academies. These institutions were specialized on, but not restricted to, basic research on the scientific frontiers. A second block consisted of the organizations responsible for higher education. The third and largest block was composed of the research organizations of the branch ministries, with the main purpose of performing industrial R&D. These organizations were coordinated by the State Planning Commission, the Gosplan, itself an instrument of the Council of Ministers, which controlled the allocation of resources. The Council was controlled by the Communist Party. This basic configuration of the S&T system of the Soviet Union remained unchanged until the demise of the Soviet Union in 1991. One of the few more extensive changes of the system between 1945 and 1991 was the creation of an institution responsible for the dissemination of innovations into the national economy. This institution, founded directly after WWII, was renamed several times during the first two decades of its existence and from 1965 to 1991 bore the name 'State Committee for Science and Technology', GKNT (see Graham 1993, pp.180).

After the Hungarian Communist's successful struggle for power, the declaration of the People's Republic of Hungary and the passing of the new constitution in 1949, the country's S&T system was reformed. As has been pointed out, the reform consisted first of all in copying of the Soviet system as it has just been laid out. The changes resulting from this reform were quite dramatic.

4.3 Actual Effects of Policies

Despite the shortcomings of the S&T system's development, the period directly following the Communist takeover in the late 1940s and the subsequent restructuring of the S&T system was a time of rapid growth for the system. Between 1949 and 1957 about 100 research institutes came into being. In the Stalinist era, until 1953, most projects were carried out in applied sciences and being used for the enhancement of industry and agriculture. 'Thus, in 1953, 66% of institute manpower, 58% of the researchers, 66% of the budget, 60% of the fixed assets and 58% of the investments for that year were for industrial research'. (Tamás 1985, p.41)

Table 4.1   R&D Manpower Growth of the Hungarian S&T System, 1953, 1958, 1961

	1953	1958	1961
Number of persons employed in R&D	18.000	22.000	29.000
Number of scientists and engineers in R&D	5.000	8.000	11.000

Source: Szakasits, D. György, 'A tudományos kutatás szerepe a gazdasági fejlödésben' (The Role of Scientific Research in Economic Development), Budapest, 1965, pp.125–128, cited in: Tamás 1985, p.40.

However, the problem with the initial growth of the institutes was that, in spite of a total growth in manpower, personnel was withdrawn from the industrial labs, where it was needed badly in a time of reconstruction. Industrial labs that were weakened in terms of manpower often gradually shrank and subsequently disappeared. As an effect, the missing in-house R&D capacity then aggravated the problems of the Hungarian economy. Without its technological base in the form of qualified personnel a firm had a hard time absorbing innovations, let alone innovate autonomously.

After the beginning of the S&T system's reorganization, the Hungarian Academy of Sciences (MTA) was soon not just a learned society, but it actually incorporated a wide variety of functions. Authors have found different ways of describing the construction of the Academy of Sciences in the People's Republic of Hungary: 'ministry of science', 'holding company operating an extensive research network' (Tamás 1985, p.40), 'ministry or governmental agency' (Székely 1988, p.1, footnote 1), 'national authority for special administration' (Darvas 1988, p.146), 'the supreme scientific body' (Tolnai/Quittner/Darvas 1985, p.115) and the like.[8] In this context it is important to notice that although the original purpose of the MTA at the early stages of reform was to do basic research, the tasks of the Academy included working on problems of the industrial sector as time went by.

The increase in personnel and funding for the MTA over the 1950s and 1960s did not only hurt the industrial labs, but also dried up the university funds and often dragged the best researchers from the universities to the MTA. Countermeasures against this development included the cooperation of MTA institutes with universities conducted under the leadership of the MTA. These cooperative programs were already begun before the establishment of the People's Republic, but were intensified after the end of Rákosi's domination of Hungary in 1954. These supplements in financial and personal resources were badly needed by the universities, because of the exploding increase in student numbers. However, they could only ameliorate the effects of the basic problem of the Hungarian university system, which before 1948 was oriented to the afore mentioned ideal of a unity of research and teaching and which now had to convert research capabilities into teaching capabilities.[9]

Similar to the arrangements in the Soviet Union, the elimination of research at the universities had found its expression in the eradication of postgraduate education at universities. As a result, the MTA had acquired another function since 1950: it granted the postgraduate degrees 'Candidate of Science' and 'Doctor of Science', comparable to 'Doktor' and 'Doktor Habilitatus' in the Austrian and German systems.

The Hungarian higher education sector in the early 1950s was deprived of many leading researchers, who had either found better working environments in the MTA or had become part of the brain drain. Again paralleling the development in the Soviet Union two decades earlier, university personnel, which was considered to be politically unreliable, was forced out of higher education institutions. Sometimes those scientists were 'hidden' by friends in jobs at the research institutes of the MTA.[10]

Scientists both at the universities and at the MTA were faced with a change in research topics. In general, the emphasis on technical and natural sciences fostered the development in these areas. In an effort to provide the economy with the technical know-how for the industrialization drive, the technical sciences were to receive more than half of all the funds for the higher education sector. The social sciences and humanities were curtailed. In addition, they felt the decoupling from the international exchange of ideas very strongly. Certain sciences were banned completely, like psychology, sociology, genetics or cybernetics. These disciplines still feel the impact of the decisions taken after WWII, even today. After all, the Institute of Sociology was founded only in 1963 and the Computer and Automation Institute in 1973, when comparable research fields had been long established in capitalist countries (Farkas 1985, p.91).

Table 4.2   Student Numbers in Hungarian Universities, 1930–1990

Year	Students
1930	12.611
1950	32.501
1960	44.585
1970	53.821
1980	64.057
1990	79.206

Source: Tamás 1985, p.54; Bessenyei/Melchior 1996, p.93.

As is shown in the table above, the student enrollment over the late 1940s and early 1950s rose by a staggering 300%. This was also due to the efforts of the Communist Party to raise the share of children of working class and farmer's families in the total student population. To facilitate this, evening and correspondence course programs were created. Furthermore, the higher education system was decentralized, with the number of higher education institutions rising from 19 in 1950 to 55 in 1955. After Stalinism, during the second half of the 1950s, however, the number of students was slowly reduced again. This measure was necessarry, since not only the higher education system had crumbled under the rapidly increasing student numbers, but there was also an overproduction especially of graduates specialized in the technical sciences, who could not be absorbed in R&D in particular, or, more generally, the economy at all.

Before the transfer of the Soviet S&T system to Hungary, industrial research had been a task of rising importance for the universities. However, after the transfer of the Soviet S&T system to Hungary the higher education sector was not only not supposed to engage itself in industrial research, but it did not have the capacities to do so.

The majority of industrial R&D in the People's Republic of Hungary was carried out in the research institutions of the branch ministries –

especially in the industrial ministry. As has already been pointed out before, the bulk of the funds going into the S&T system was earmarked for industrial research. The branch ministry's institutes were among the main beneficiaries of this growth. Between 1949 and 1957 an astonishing amount of 70 research institutes were established in Hungary, of which about 40 engaged in industrial R&D.[11]

Industry, however, did not benefit much from the sudden growth of the industrial R&D system. Serious technology transfer problems were arising from the fact that the institutions performing the R&D were not directly attached to the companies.[12] Moreover, the existing in-house R&D capabilities of industrial companies were neglected. Finally, the average qualification of the industrial R&D system's personnel was low.[13] A mere four percent of all Hungarian researchers holding the 'Candidate of Science' degree, which is comparable to the United States' 'Ph.D'. or the German 'Doktor', worked in Hungarian company R&D units in 1991.[14]

At the time of Stalin's death the Hungarian innovation system had all the features of a typical Soviet system as it has been described by Dimitry Piskunov and Boris Saltykov:[15]

- it is organized around sectors, therefore allowing for a great amount of isolation between the fields and creating barriers for interdisciplinary discourse;
- management and control are centralized; institutions such as the Academy of Sciences have a multitude of tasks to fulfill;
- planning and management of science are dominated by the fulfillment of the plan, an idea lurking in the papers in the form of formal indicators;
- the allocation of input is rigidly rationed through the plans.

The high degree of centralization of the S&T system's control functions was another factor hindering the development of the S&T base of the country enormously. The political control of S&T was especially strong in the years of Stalinism, when it was administered by Mátyás Rákosi and Ernö Gerö, who was responsible for science and economic policies. The political pressure on scientists not only made it impossible to freely follow research lines, which were judged as promising or to get into contact with researchers working in the same field, but even went so far that the publishing of the results of one's own research was screened, if one happened to work in a strategically important field. The compartmentalization of the different research fields was pushed to an extreme, when the results of the scientist's work were channeled to the top echelons of power and disseminated from there only when it seemed opportune. This development was the all-too-real manifestation of the truism that 'knowledge is power' (Wissen ist Macht), well known in both Hungary and German speaking countries.

A specifically striking example concerning the control of the information flow within the S&T system were the policies regulating statistical information and economic research. In post-WWII Hungary, as in every modern society, the knowledge about the economic processes was quintessential for policy planning functions as well as further scientific analysis in a number of disciplines. However, the only Statistical Yearbook published between 1946 and 1953 was printed in 1948 and covered the years from 1943 – 1946. Table 4.3 contains information on the Yearbooks produced between 1949 and 1953.

Table 4.3   Statistical Yearbooks during Stalinism in Hungary, 1949–1952

Date of Publication	Year Covered	Volume of Numbered Copies	Official Classification
1949	1947	400	Confidential-Manuscript
1950	1948	400	Strictly Confidential
1950	1949	400	Strictly Confidential
1951	1950	250	Strictly Confidential
1952	1951	150	Strictly Confidential

Source: Péteri 1993, p.151.

Between fifty and sixty persons were regularly provided with the information contained in the Yearbook between 1949 and 1956. Not only the information itself, but also the list of those who could obtain the statistical information was considered 'top secret'. The regular publication of statistical information in post-WWII Hungary started only in 1957, when a yearbook was published that covered the period from 1946 – 1955.[16]

4.4 Possible Explanations for the Differences

The results of an analysis of the Hungarian S&T system's transition from a form of capitalism still bearing the imprints of a war economy in 1945 to realsocialism are twofold. On the one hand, the realsocialist organization of S&T made advances possible in those fields, in which the strong and repressive usage of the factors of production were key to success. In the case of Hungary one such example is atomic energy, in the case of the entire Council for Mutual Economic Assistance (COMECON) such examples might be space research or weapons technology. In these fields the realsocialist countries, including Hungary, could catch up with the West and at times even attain a formidable lead over the Western countries, as was the case in space research. Furthermore, in a number of basic sciences as mathematics, chemistry or theoretical physics, in which Hungary (and Russia) were already strong before realsocialism, the position of the realsocialist countries in science stood its ground.

On the other hand, the same factors leading to success in large scale S&T projects – 'big S&T' – led to a further lagging behind in areas, where small and at times cooperative projects in a competitive environment were helping the development of certain technologies. Here, classic examples are computer and telecommunication technologies.[17] In these new technologies the centrally planned economies, including Hungary, were lagging behind and were not capable of catching up, despite all efforts. What are the specific reasons for this backwardness in a number of fields? And why did at times good and innovative research not find its way into the market?

One explanation for this backwardness can be found in the tendency to avoid risks throughout centrally planned economies. Innovative behavior and the risks attached to it always have been circumvented in Soviet-type economies. A major reason for this risk aversion, besides the afore-mentioned organizational problems of compartmentalized and centralized hierarchies, is the inhumane way in which industrialization was pushed and in which the technical intelligentsia was marginalized especially in centrally planned economies under the leadership of Josef Stalin (Graham 1993b) and Mao-Tse-Tung (Wang 1993, pp. 46).[18] Obviously, the regimes were also opposed to one of the preconditions for innovations, namely the free dissemination of information, which makes competition of ideas, innovations and therefore progress itself possible in other systems. The bad information infrastructure in the Eastern Bloc was not only merely the result of a weak technological basis in information technologies, moreover, it had its political purposes, too (see Skolnikoff 1993, pp.96). This can be seen in such measures as the keeping control of the number of type-writers and, later, copiers and computers in usage in the COMECON.

This opposition to innovations could not only be found in bureaucracies, but also in those parts of society that in capitalist systems are innovative almost by nature – industry. The following statement of a Hungarian is typical of a centrally planned economy, even if it had been through that many stages of reform as the one in discussion had: 'No management has ever been fired for missing an opportunity, but failed innovations often have been held against them [the managers, PB]' (Geipel 1991, p.212). Innovations are feared in centrally planned economies for their uncertainties and often disruptive nature.

In addition to the potential insecurities embodied in innovations, there is often a lack of incentive structures enhancing the willingness of innovators, be it researchers or institutions, to make their R&D results public. However, this has been less of a problem in Hungary, where since the late 1950s this lack of incentives has been discussed and was subject to reforms that will be described later. Still, in the absence of property rights, personal as well as intellectual, innovators chose to guard their 'advantages' against competitors. As a result the jealously-guarded inventions often had to be reinvented again and again.

This, of course, is all the more true for technologies perceived as strategic, i.e. where import substitution strategies – efforts to supplement imports with indigenously produced goods – are implemented. All the sheltered centrally planned economies suffered twofold in these areas. First because technologies were not bought from the sometimes far superior Western producers, but were copied with considerable time lag, often through reverse engineering. Second, because there was practically no linkage between military and civilian R&D, in the former sector available technologies had to be reinvented in the latter sector because of perceived security risks. Spin-off effects, i.e. cases where military inventions were introduced to civilian markets, under these circumstances were known to only rarely happen (Geipel 1991, p.215, and Chiang 1990, pp.399).

While the autarchic tendencies in Hungarian industrial and S&T policy making were strong during the first decade of the People's Republic, they receded over time. Indeed, in subsequent years Hungary served as a technological gate keeper in the iron curtain for the Council for Mutual

Economic Assistance, COMECON. Hungary was used for the transfer of technology from the West to the East.[19] As a consequence, it was sometimes easier for Hungary to overcome the technological constraints of the centrally planned system than for other, more autarchic, national economies of the Eastern Bloc.

But not only military and civilian R&D are kept apart in Soviet style economies. Worse still, basic and applied science as well as engineering, respectively, are split up, too. The separation of scientific research and higher education, and the segregation of R&D and the economy have been described as 'detrimental to science' (Pungor/Nyiri 1993, pp.25–39). The division of R&D on the one side and the economy on the other side has proven to be among the fatal consequences of different features already mentioned. In this respect the sectorization of the innovation system, the funding of institutions rather than projects, the missing incentives for sharing inventions as well as the general opposition towards innovation in the culture of these systems are characteristic features.

A term which stands for the problems resulting from a number of typical features of the realsocialist system is the term 'quasi-development'.[20] 'Quasi-development' means

> ...that in our (countries) everything is only 'almost' working, only almost adequate, that we only almost command the competence (needed), that we have only almost all the experience necessary ... Think of all that fills from week to week the pages of our comic magazines but which is ignored when the assessments of our national incomes and living standards are made. These are the phenomena of a 'quasi-developed' structure and 'not' of a low level of development.[21]

For the S&T system 'quasi-development' translates into phenomena such as the existence of a large number of researchers in comparison to the size of the population, but bad equipment and small libraries for these scientists. It means large sums of money spent on S&T, but no clear funding structure and, subsequently, no steering possibility for policy-makers. It also means R&D results having little or no effect on production. It means missing possibilities for researchers to communicate with their peers in other countries. To put it in a nutshell, the Hungarian S&T system after its transition to realsocialism can, in a number of areas, be seen as 'quasi-developed'.

Technological change, one of the basic requirements for a thriving economy, was hampered in the People's Republic of Hungary by all the factors discussed here.[22] Under these circumstances the 'science-push' notion, the model underlying the Soviet S&T system, could not to be successfully implemented.[23] Recognizing these problems, a group within the Hungarian Communist Party gathered around the economist Rezsö Nyers and started to work on the concept of a reform of the Hungarian economic system under the auspicies of Janos Kádár.

Notes

1   Judith Mosoni gives a total size of 38 research institutes and about 400 researchers for the Hungarian S&T system in 1938. This number, however, seems to exclude the university system. See Mosoni 1994, p.1.

2   For an interesting analysis of the cleavages in the Hungarian science system and the events in the aftermath of WWII, see Péteri 1993.

3   This is very similar to what the Soviet Union went through in the 1920s. After the original plans had been to close the Russian Academy of Sciences, policy-makers had to concede that there was no way to 'transform a good Communist factory worker into an internationally known physicist' (Graham 1993, p.177).

4   For an account of the MTA's transition from its original role in society to its new role in the S&T system of the People's Republic of Hungary see Petéri 1993.

5   This was the second round of expellations from the MTA in only 7 years. In the Stalinist purges of 1949 over 100 researchers had been expelled (Kosáry 1991, p.230). Researchers subject to both the 1949 and 1956 events often were readmitted only in the 1980s. Domokos Kosáry, the late president of the MTA of the time after the transition to capitalism, was a good example for these events: he had been expelled in the 1950s and was readmitted in the 1980s.

6   Péteri's translation from the Hungarian edition of J.V.Sztálin, 'A Leninizmus Kérdései', Szikra, 1953, pp.410, in: Péteri 1994.

7   For a more detailed description of how the Soviet science system was constructed in the 1920s and 1930s, see Graham 1993, pp.173.

8   However, one interview partner knowing the MTA intimately insisted that the influence of the MTA was widely overstated and that the Academy played a lesser role in the S&T system of the People's Republic than is normally believed.

9   This problem is reminiscent of the difficulties the Western European university systems were confronted with in the 1970s and 1980s, when the number of students grew rapidly, due to – mostly Social Democratic – educational policies. Research capacities were converted into teaching capacities and all the measures taken by the state were dwarfed by the increase of student numbers. As a result, research is increasingly being excluded from Western European universities.

10   Several interview partners at the MTA pointed out that a number of ex-university personnel was quietly kept on the payroll of the MTA, working in areas less visible to the public.

11   Aba, Iván, 'Múszaki-tudományos kutatás Magyarországon' (Scientific-technological research in Hungary), Múszaki Könyvkiadó, Budapest, 1965, cited in: Mosoni 1994, p.2.

12   Loren Graham states that in the Soviet Union the isolation of research and industry was identified as problematic as early as 1925 (during the New Economic Policy of Lenin). See Graham 1993, p. 179.

13   Several Hungarian interview partners confirmed that the average education of the researchers in industrial R&D was lower than in the MTA. Moreover, a minority of interview partners suggested that the national economy will profit from the laying off of the larger part of the industrial R&D personnel they deemed to be unproductive in many cases.

14   Pécsi 1993, p. 27; in the Soviet Union the situation was similar: Graham states that in 1982 a mere three percent of all Soviet researchers holding the Candidate of Science degree worked in industry (Graham 1993, p.181).

15   See Piskunov/Saltykov 1992: Saltykov was then the minister for science, higher education and technology policy in Russia.

16   The information on the handling of statistical information during Stalinism is drawn from the detailed description of the policies in Péteri 1993.

17   For an account of computer and information technology policies, see Geipel 1991.

18   Wang's and Graham's accounts are not devalued by the observation that many high ranking officials in realsocialist systems were educated as engineers, because they changed their societal function when becoming a party cadre. Nor are they made obsolete by the fact that Communist strategies included the use of the technical intelligentsia for the take-over of countries, as is described by Péteri 1993. In the latter case the technical intelligentsia and its longing for acceptance in society and academic institutions was simply used for power politics.

19   A Hungarian interview partner in a senior position pointed out that it was almost always possible for Hungary to acquire any given technology from the West, regardless if the technology was on the COCOM (Coordinating

Committee for Export to Communist Area) list or freely available. However, the financial costs involved in getting a hold of technologies on the list sometimes were exceedingly high.

20   The term stems from Jánossy, Ferenc, who, during the economic reforms in 1969, wrote an article, in which he scourned the shortcomings of centralized planning. His article has been partially translated and cited by Péteri, György 1994, p.20. See the original: Jánossy, Ferenc, 'Gazdaságunk mai ellentmondásainak eredete és felszámolásuk útja' (The Origins of the Present Economic Problems of Hungary and the Way to Solve Them), in: 'Közgazdasági Szemle', July-August 1969, pp. 806–829.

21   Jánossy 1969, p.828, in: Péteri 1994, p.22.

22   For a general statement on the key problems centrally planned economies have with the processes of technological change, see Skolnikoff 1993, pp. 128.

23   While it is true that the 'science push' policy paradigm was abandoned in the capitalist systemsduring the 1970s, too, S&T planners there still had more success in applying the idea to the real world than in realsocialist systems.

5 The Second Transition: The New Economic Mechanism

5.1 Historical Development of the S&T System

After a period of rapid economic growth during the first half of the 1950s, the Hungarian economy slowly cooled down over the next decade. At the end of the second five-year plan, in the late 1950s, it was becoming increasingly clear that serious problems were hampering the economic development of the country. Despite a still substantial economic growth, the usage of the factors of production that had been enforced had put a lot of stress on the available resource base. This was specifically the case for labor, capital and technology.

On the one hand, labor and capital, both drawn from agriculture, were becoming scarce resources. The labor reserves of the agricultural sector were diminishing and a large share of women was already included in the labor force (Marer 1986, p.65). Capital was missing, too, as the Hungarian exports to the capitalist countries were shrinking – which was amongst other things a result of the weak technological basis.

Technology, on the other hand, had been scarce right from the start. As a result the quality of products was lower than in capitalist Western Europe and could be sold only on COMECON (Council for Mutual Economic Assistance) markets. An example for this is a report of the Hungarian embassy in Austria, which was quite critical about the Hungarian tool-making machines' chances on the Austrian market. First, the embassy reasoned, Hungarian tool-making machines had a technologically low standard and were therefore less demanded in Austria and second, Hungarian producers could neither meet the delivery times, nor the extensive service of Austrian or Western European competitors.[1]

Industrial productivity is a measure for the technological level of advance of a country. A comparison of industrial productivity in Hungary, Austria, Czechoslovakia and France for the years 1967–1969 shows a considerable lack of productivity for almost all sectors of Hungarian production. The technological backwardness of Hungary was especially pronounced in sectors where development began later, including practically all heavy industries. A higher level of Hungarian industrial development can be seen in other sectors, mostly in light industries and specifically in foodstuff production.

Table 5.1   Industrial Productivity in Hungary, France, Czechoslovakia and Austria, 1967–1969

Hungary = 100

Sector	France	Czechoslovakia	Austria
Mining	320	351	254
Food	134	115	85
Textile, Clothing	139	133	115
Metallurgy	203	147	125
Machine Building	289	147	152
Industry Total	208	166	140

Source: Marer 1986, p.66.

For the Hungarian S&T planners it became increasingly clear that the old institutional structure of the system was inadequate. This observation led to a series of innovations regarding functions and structures of the S&T system. Already in 1959 each company was required to set up a Technical Development Fund (MÜFA, Müszaki-Fejlesztési Alap), out of which its R&D expenses were to be paid. A year later, in 1960, the first National Long Range Plan for Scientific Research (OTTKT) was approved by the Council of Ministers. It was a gigantic effort to plan the national S&T effort, including 103 main goals and 23 research guide lines for the different fields of science and engineering. The OTTKT was troubled from the start on. Its main deficiencies were the inability to differentiate between the needs and possibilities of the different disciplines as well as the fact that the plan was increasingly linked to the economic policies (Müller 1985, p.225). Finally, in 1961 the State Committee for Technical Development (OMFB) was created by the Council of Ministers. The committee's task was to work out plans for policies concerning critical technologies. It provided advice to other institutions, first of all to the powerful Council of Ministers. OMFB served also to facilitate interest harmonization in the S&T sphere (Tolnai/Quittner/Darvas 1985, p.116).

The 'Science Policy Guidelines of the Central Committee of the Hungarian Socialist Worker's Party', the Magyar Szocialista Munkáspart (MSZMP), as the Communist Party had named itself, were set up in 1969 and revised later, in 1977 and 1985, after a long period of discussion within the party and with outside experts. By naming them 'Guidelines' the Central Committee made clear that it did not want to interfere with the freedom of science and the arts – a statement that is obviously only of relative truth, given that the funding of arts and sciences was strongly dependent on the content of the planned projects.[2] The Agitation and Propaganda Committee of the Communist Party's Central Committee in particular has been charged in recent years with interpreting the 'Guidelines' vow to respect the freedom of science in a conveniently loose fashion.[3] Nevertheless, the degree of freedom was certainly higher than in other realsocialist Eastern European systems.[4]

The 'Science Policy Guidelines' also prescribed the institutionalization of the Science Policy Committee (TBP). It consisted of the deputy prime minister, who was in charge of science policy, the ministers of major branch ministries featuring research institutes, the President and the Secretary General of the Academy of Sciences (MTA), the Chairman of the State Committee for Technological Development (OMFB), the President of the National Planning Office and the Minister of Finance. This committee was the highest institution for the implementation of science policy, which passed resolutions that were of binding character for all governmental agencies. Before the resolutions were passed, they had to be agreed to by the Council of Ministers, the inner circle of power in the pragmatic and less party-oriented Kádár regime. In fact the committee was dominated by the MTA and the OMFB, which was responsible for 'in most cases the preparation and coordination of government-level decisions – in most cases their implementation and national coordination – and the presentation of reports to the government' (Tolnai/Quittner/Darvas 1985, p.115).

After the higher education sector had been shrinking in the second half of the 1950s, the number of students and different forms of institutions were expanding again in the 1960s and 1970s. One visible reaction to the perceived needs of the economy was the creation of technical colleges, which were not granted the status of universities. The total number of students during the first half of the 1960s came close to 100.000 – a number which was not increased decidedly until the transition to capitalism began in the late 1980s (Bessenyei/Melchior 1996, pp.51).

The S&T system was not the only societal subsystem under reconstruction. On May 7, 1966, the Central Committee of the Communist Party produced a resolution that led to a basic economic reform – the New Economic Mechanism (NEM) – finally launched in 1968. The principle of central planning was to be replaced by what came to be called an indirect bureaucratic system. Free market elements were introduced to the economy of Hungary for the first time in two decades. Management and workers took part in a profit-sharing system, prices reflected to a certain extent the valuation of the products by the market. In 1968, as an effect of the NEM, the central determination of prices was abandoned for 12% of agricultural products, for 28% of domestically produced materials and semi-finished goods, for 78% of industrial products and for 23% of consumer goods and services (Balassa 1982, p.6/7).

In the following years, the reform's scope was widened. The central allocation of materials was dismantled and the national economy opened up towards the international environment. One immediate effect of the boom which followed the introduction of market elements was an increase in the number of researchers employed in the company R&D sector. In addition, during the NEM many S&T institutions were forced to obtain funding through external R&D projects not funded directly by the state. This, of course, was an attempt to strengthen the relationships between the realms of industry and academia.

In 1972 the economic policies began to be more cautious and moved backwards into the direction of a recentralization – as an effect of the concerns within the left wing of the Communist Party about rising social cleavages due to income differences. When the economy experienced serious problems during the mid 1970s, however, the reaction of the government was to slowly change course again. Beginning with 1977 one can speak of a reinvigoration of the NEM.

Despite all the successes of the NEM, it did not eliminate some of the underlying problems of the Soviet style system. For the economy among one of the most fundamental problems was the soft budget constraints the enterprises faced – there was practically no possibility of bankruptcy for firms, even in the case of repeated loss making – and the strength of the indirect system of bureaucratic control that would sometimes intervene quite directly (Hanson/Pavitt 1987, p.35). The soft budget constraints were prevalent until the beginning of the transition back to capitalism in the late 1980s.

In the 1980s a series of changes edged the national economy even nearer to a market system. A development crucial for the S&T system was that inventors could, for the first time, obtain property rights for their successfully introduced innovations in the 1980s, including the right to earn royalties from patents. Moreover, the banking system was reformed and the national bank split into several business units. Small businesses flourished, cooperatives were allowed to use state enterprise infrastructure and the strict plan system was gradually removed. Lastly, strategic technological and economic goals were written down in the National Medium Term R&D Plan (OKKFT) in 1981.

5.2 Policy Maker's Models and Intentions

In 1977 János Kádár could quite rightly point out in his article 'For Socialism and Peace' that in comparison to 1938 figures Hungarian industrial production had increased 1000%, agricultural production 150% and the GDP 500% respectively. However, his prediction that the country would reach the level of the most advanced industrial societies during the next 10–15 years proved to be incorrect (Kádár 1984, 27/28). Nevertheless, in the early years of the New Economic Mechanism (NEM) Hungary's catching-up with Western Europe seemed doable.

One of the ways the NEM wanted to facilitate this rapid economic development was the integration of industrial research and industry. There were many attempts in centrally planned economies to overcome the problem of separation of industrial research and industry. The Hungarian NEM was by far not the only one that tried to do so. Loren Graham finds that in the case of the Soviet Union

> an examination of the record of resolutions passed by top Communist Party and government organs shows that in 1926, 1929, 1947, 1961, 1965, 1966, 1975, and 1985 a major goal of the directives was bringing science and production into closer contact. (Graham 1993, p.182)

However, not one of the resolutions listed here came close to the extensive scope of the Hungarian NEM. The NEM certainly was the most radical reform program ever implemented in the Eastern Bloc.

During the Rákosi era the funding of the S&T system was taken care of by the state. From the mid 1950s on, however, the funding for the whole S&T system was not centrally allocated by the state any more. In doing so, an effort was made to overcome the distortions and inflexibility of the former funding system. Consequently, university departments and independent research institutes were allowed to undertake research ventures. The financing of these ventures was, however, still provided by the state.

In 1959, still well before the NEM, the guidance of the S&T system by the bureaucratic system was reduced further. As a result, the development of products in industry was less dependent on the state plans. Moreover, each company had to form a Technical Development Fund (MÜFA, Müszaki-Fejlesztési Alap), out of which its R&D expenses were to be paid. The state, on the other hand, still took care of basic research and research performed by the institutes of the specialized ministries. The creation of the Central Technical Development Fund, the KMÜFA, funded by the MÜFAs, was part of a reform in 1961 which set up the organization that was to disperse the funds of KMÜFA, namely the afore-mentioned National Committee for Technical Development (OMFB). The KMÜFA was financed by a sales tax on companies – more precisely a percentage of the sales minus the direct material costs of production. Depending on the sector, the percentage of the sales which was the basis for the tax, varied, ranging from 2% to 12%.

The Higher Education Act of 1961 brought a change in policies affecting university research. The law stated that university teachers have to use part of their time doing research. With this regulation, the emphasis in higher education on the teaching of the principles of Marxism-Leninism was significantly reduced. The criteria for the selection of teaching personnel were, at least officially, changed from the degree of political reliability to the scientific competence of the person in question. Since then the amount of scientific research performed within the university system has been on a slow but steady rise (Erdey-Grúz/Kulcsár 1975, p.23). Most of this research was basic science, but with the implementation of the NEM universities were increasingly engaged in applied research, too.

The document most aptly describing the reasoning behind the reforms of the S&T system is the same document, which was the very basis for the reforms: the 'Science Policy Guidelines of the Central Committee of the Hungarian Socialist Workers' Party'. The 'Science Policy Guidelines' gives an overview over the condition of science in the country. The interrelations between societal and economic goals and R&D are described: science, according to the document, not only affects the social consciousness, but also enhances the skills of people and forms a basis for the management and organization of production. 'The achievements of science are utilized in the interests of the entire society'.[5] The document sees science as a problem solver, which is typical for the time and for the 'demand pull' policy paradigm.

Moreover, in the 'Science Policy Guidelines' the structures of the Hungarian S&T system are criticized. It goes on to suggest that both the selection of research topics and the utilization of the results of R&D, should be enhanced by the creation of a modern management system of science. More specifically, the fact that Hungarian research concentrates on a small number of disciplines and topics is attacked in the paper. Other problem areas identified in the document are

> the fact that the forces of science are frittered away as a result of the autarkist policy introduced in 1945, which disregarded the natural endowments and the real capabilities of the country; certain traditional attitudes which consider knowledge important for its own sake; inadequate auxiliary manpower is provided for researchers; the level of research centres is unevenly distributed over the country, etc. (Vas-Zoltán 1985, p.283)

It is important to understand the 'Science Policy Guidelines' as a highly political document. It criticizes directly the policies, on which the Soviet style organization of the S&T system were based at the time.

Two obvious effects of the 'Science Policy Guidelines' are the reform of the Academy of Sciences (MTA) in 1970 and the 'rehabilitation' of the universities, which were henceforth granted more possibilities of income (industrial R&D) and autarchy. Generally the S&T system was liberalized, the degree of freedom was enlarged[6] and the structures and to some degree also the functions of the system moved into the direction of the organization of S&T in capitalist economies.

From 1972/73 onwards independent institutes for industrial research had to operate as profit-oriented businesses. The MTA and other institutions of the S&T system, as for instance university departments, were given the opportunity to contract research out.[7] They were even forced to do so, because funding by the state was diminishing ever since the start of the NEM (Csöndes 1985, p.188). Moreover, not only was the state funding for

R&D held steady in absolute numbers, but also was en bloc institutional funding decreased and replaced by project funding. The intention of the policy-makers was to create an incentive for researchers to engage in working relationships with industry and to thereby bridge the gap between R&D and industry.

The companies were pressed to increase the quality of their products and to adopt innovations more quickly. The policy maker's wanted to foster exports to market economies in order to be able to import technology in the form of capital goods such as machinery. The overarching long-term goal of government was, again, the closing of the technological and, subsequently, also the national income gap to the West[8] – which strongly reminds of the goal of earlier periods, both the one following the 'Ausgleich' and the other after the end of WWII.

One tool for the furthering of these goals was the fostering of foreign trade. Trade with non-COMECON countries until the introduction of the NEM had been understood as necessary in order to raise convertible currencies from Western countries. Now, trade with capitalist countries was seen as the possibility to gain new markets and new technology. For this strategy Austria had been assigned a special role by Hungarian policy makers. Firstly, the Western neighbor was technologically less advanced than other European capitalist countries. Therefore, the policy makers hoped that imports from Austria would not disturb the already strained trade balance with other Western European countries. Secondly, Austria was seen as a possible test case for the export of Hungarian products to other Western countries. Since Austria was economically less developed than the larger part of the Western countries, one assumed it would be easier to try to set foot on this market than on others. Thirdly, since the technological development of Austria was almost comparable to that of Hungary, the differences between capitalist and realsocialist systems would be less visible for the Hungarians in a cooperation with Austria than, say, in a cooperation with the FRG.[9]

In the late 1970s, in order to open up further possibilities of funding and to enhance the cooperation between the R&D institutions and industry, it was made possible to transform research institutions into economic associations. This transformation was possible for organizations whether or not they were engaged mainly in basic or applied science, development, engineering or actual production. In case MTA institutes were involved, the Academy defined these constructions as 'associations of...members of the

Academy's institutes who with the consent of...their institutes carry out business-type activities...in their subject field/s' (MTA 1988, p.54). The new economic associations could function in only one institute or as a cooperation of either many institutes or institutes and enterprises. The first economic associations with MTA institutes were formed in 1978. In late 1983 twenty-two such entities existed. Institutes also formed joint stock companies from 1984 on.

Clearly, the idea behind all these policies was the 'demand-pull' or 'science-pull model' as laid out in the introduction. The NEM should create a 'simulated market', with the goal of arriving at a 'regulated market economy' (Marer 1986, pp.18) or a 'market socialism'.[10] The price mechanisms of the simulated market were supposed to have a regulative function, steering supply and demand, but were at the same time held under control by the state bureaucracy. Analog to the workings of a market economy, the demand of the market for new products should lead to a demand for innovation, which would be met by the industrial R&D system.[11] The R&D institutes now could choose from a variety of incomes: the shrinking en bloc funding by the state, government procurement, national technical development programs administered by OMFB and, finally, the MÜFAs of the companies which provided an increasing source of income.

5.3 Actual Effects of Policies

Between 1967 and 1973 the Hungarian economy's average annual rate of growth was 6.2%. Even better, the cumulated growth rate of total factor productivity in the years 1962–1967 doubled in the years 1967–1972. Availability, assortment and quality of goods and services improved.[12] The already extensive national R&D expenditures were expanded considerably, too, as can be inferred from table 5.2.

**Table 5.2   R&D Expenditure as a Percentage of GDP, 1971–1981,
at Current Prices**

Year	Expenditure as a Percentage of GDP
1971	2.48
1973	2.55
1975	2.89
1977	3.07
1979	3.01
1981	3.00

Source: Csöndes 1985, p.183.

From this time on criticism was brought forward against the rising expenditures on R&D during the 1970s, which was a time of stagnating growth.[13] After all, the steady growth which was to be found at the beginning of the NEM was now facing a negative influence by a world-wide recession, which had come about due to the first oil-crisis in 1973, and by the deterioration of the terms of trade with the capitalist countries as well as the Soviet Union.[14]

One positive, although not long lasting, effect of the NEM was the growth of the number of researchers working in enterprises and the improvement in the quality of the work done. The number of people in the area of enterprise R&D grew from 17.600 in 1967 to 24.867 in 1971, or from 31.9% to 35.2% of the nation-wide R&D staff, respectively (Tamás 1985, p.45).

The economic policies began to get more cautious in 1972 and subsequently moved backwards in the direction of a recentralization. Again, the impact on the S&T system could be felt immediately. R&D capacities in industry shrank and the ratio of researchers in industry to researchers on a national scale came down again from the 35.2% in 1971 to 31.9% in 1975, exactly the percentage of 1967. The increase of the total number of people working in R&D first slowed down, then, from 1975 on, the number even began to decline. The reinvigoration of the NEM in 1977 did not lead to a renewed increase in the number of researchers in the industrial R&D system. The slower growth of the national economy and the reduction of funding of R&D might be among the reasons for this stagnation.

As a result of the creation of the afore-mentioned Company Technical Development Funds (MÜFAs) in the 1960s, companies spent more on and performed more R&D than before the reforms were introduced. Companies' R&D was larger than the other sectors of the S&T system: 58.2% of the total staff and 67.5% of the total expenditure on Hungarian R&D in 1981 can be assigned to the companies. Table 5.3 reflects the percentages of staff and expenditure of Hungarian R&D units in 1981 according to sectors.

Table 5.3   Hungarian R&D Units According to Sectors in 1981
(Percentage)

Sector	Staff		Expenditure	
Companies	37.7		38.6	
R&D Units Serving Companies	20.5		28.9	
Higher Education	19.3		11.2	
Public Services	22.5		21.3	
of these: MTA		11.8		14.0
Public Health, Culture etc.		10.7		7.3
Total	100.0		100.0	

Source: Csöndes 1985, p.187.

This domination of companies' R&D over the other sectors is also reflected in the shares of R&D expenditure according to governmental institutions. The Ministry of Industry easily outspent the other organizations, as can be inferred from table 5.4.

Table 5.4   Distribution of Hungarian R&D Expenditure According to Sectors in 1980

Sectors	Percentage
Ministry of Industry	51.7
Ministry of Education and MTA	21.5
Ministry of Agriculture and Food	11.3

Source: Farkas 1985, p.102.

Again, the domination of industrial R&D in the Hungarian S&T system is reflected in the expenditure on R&D in the various research fields. Table 5.5 shows the distribution of R&D according to fields: the technical sciences, which can be utilized most easily for commercial purposes, dominate over the other disciplines by a large margin.

Table 5.5   Breakdown of Total Hungarian R&D Expenditure in 1975 and 1980 (in Percentage)

Field	1975	1980
Technical Sciences	64.8	64.6
Natural Sciences	13.5	16.2
Agriculture and Forestry	11.5	10.3
Social Sciences and Humanities	6.9	5.4
Medical Sciences	3.2	3.5

Source: Farkas 1985, p.102.

The reforms in the allocation of funding, which partially predated and partially originated from the NEM, had different effects on the structure of interest of R&D institutes on the one hand and industry on the other. This was due to the fact that the interests of the two sides involved were in some respects complementary.

While the R&D institutes had an interest in seeing their work implemented, the companies did not. The reasons for the companies' disinterested behavior were the same as for their neglect of innovation during the unreformed centrally planned economy which has been described before: namely, that, in a nutshell, the companies even in the times of the NEM feared the disruptive nature of innovations.

As a result, project reports from R&D institutes were frequently shelved by the company management. In the cases of successful implementation of research, the researchers' personal driving force was often the reason for the innovation and not the company's responsiveness to a perceived need to innovate (Balázs 1993, p. 543).

While the policies had made the industrial R&D system more responsive to the needs of industry – although this did not change companies' lack of willingness to innovate – it had esoteric side-effects, too. Not only were the institutes regularly underfunded, but the salaries of researchers were low in general. As an effect, each researcher tried to engage in a large number of projects, often not spending much time on each one. Moreover, each person heading a project could contract out part of the project to another institution – and thereby not only enhance a supposed 'interdisciplinarity' and thereby raise the chances to get funding, but also enlarge his or her personal network and begin a relationship that ensured the possibility to be included in somebody else's project.

> The low wages therefore stimulate researchers to support and finance each other by means of orders and contracts. Thus, a large part of funding (often more than half) for any project flows out in the form of personal income, contradicting the intentions both of the R&D management and of the prevailing economic policy. (Balázs/Hare/Oakey 1990, p.733)

The strategy of government to force R&D institutions to contract out had another rather unexpected side-effect. Even institutes which formerly had been primarily engaged in basic research, such as the institutes of the Academy of Sciences (MTA), had – and until today have still have – to acquire up to 50% of their income through contracting out.[15] The situation was aggravated by the fact that up to the latest transition from realsocialism to capitalism there were no non-profit organizations in the country to provide additional funding. As a consequence, the research institutes gradually became more and more engaged in applied research and in activities that included the production and marketing of consumer products of all kinds. Sometimes these activities were totally unrelated to the original research specialization or research at all[16] and thereby the institutions slowly lost their R&D capabilities.

As a result, significant changes in the activities of the institutes of the MTA occurred. Table 5.6 shows that the ratio of basic research declined sharply, whereas activities such as services for industry and production of goods were rising. Consequently, hand in hand with the NEM came a slowly increasing number of research institutions of the MTA which practically became part of the industrial R&D system. The share of basic research in total R&D dropped from 15% in the 1970s to 9.7% in 1984 (Darvas 1988, p.151).

Table 5.6   Percentage of Types of Research Done of the MTA, Calculated on the Basis of Expenditures on Research Topics, 1970–1985

	Basic Research	Applied Research	Experimental Development
1970	58.6	28.4	13.0
1975	49.2	27.3	23.5
1980	45.6	37.0	17.4
1985	41.0	45.0	14.0

Source: MTA 1988, p.50, p.62.

Moreover, the distinction between mainly applied and basic research oriented institutes was becoming more and more blurred. The very policies that encouraged contracting out and which were meant to be beneficial to research and industry, turned out to hinder the progress of both. 'The originally different types of institution were homogenized and atomized: most had the same mixture of activities, while their 'business' activities were based on small groups'.[17]

Nevertheless, it should be noted that not all institutes of the S&T system were suffering to the same extent and that the above mentioned deviations did not occur throughout all the years since the invigoration of the NEM. For example, all disciplines represented in the MTA grew until 1977 and particularly so in the first half of the 1970s. From 1970 to 1980 the total MTA expenditure grew by an annual rate of 13%, much faster than the R&D expenditures of the whole S&T system. The growth occurred unevenly, however. While the technical disciplines in terms of researchers grew by 82% from 1970 to 1977 and the natural sciences by 64%, the social sciences only grew by 16% and the agricultural sciences expanded by a mere 8%.[18] The MTA itself offers an explanation for this development: It is the better income possibility of the former disciplines that made them fitter for growth.[19] The income possibilities the scientists had found are accounted for in the following way:

[i]n addition to production-related special research tasks, commissions from enterprises might extend to so-called scientific services (measurements,

expertise, computations, etc.) or to turn out unique or small-series products requiring special expertise or equipment. (MTA 1988, p.45)

Apart from the MTA, a second sector of the Hungarian S&T system suffered comparatively little from the deterioration in funding: the university system. As has been established, since the early 1950s the university system has been growing in terms of budget, teachers and students. Since the 1960s the amount of research performed by universities has been growing, too. In 1985 a startling 937 R&D units existed in the higher education sector. It should be mentioned that the large number of units is not so much a result of frantic R&D activity, but more of the large number of total higher education units, many of which are small and consist of only one faculty or department.

In 1985 the R&D expenditures of the university research units amounted to HUF 3 billion, or 12% of the national R&D effort (Darvas 1988, pp.169). Much of the research performed had little direct influence on industry as it was basic science oriented. However, as the cooperation between MEFI, an agricultural machinery development company – now privatized – and the Budapest Technical University shows, there are notable exceptions to this rule. The cooperation has a decade long history and has led to technical developments in the field of agricultural machinery.[20]

During the 1980s, budgetary restraints in the form of a huge national debt, which had accumulated in the aftermath of the first oil shock, made a more active governmental funding policy impossible. To ameliorate the worst effects on the country's basic research system the National Research Foundation (OTKA) was set up in 1985.[21] It awarded grants for basic research projects on a competitive basis. Between 1986 and 1990 around 3000 applications were received. 1200 projects were granted financial support, which amounted to a total of 4 billion Forint (Pungor 1991, p.23). The fund was administered by the MTA until January 1, 1991, when it was set up as an independent legal entity. Funding of applied research was still administered through the KMÜFA, which had been of diminishing importance during the last years of the People's Republic.

5.4 Possible Explanations for the Differences

The Hungarian New Economic Mechanism (NEM), certainly the most fascinating reform project of a COMECON (Council for the Mutual Economic Assistance) country, was successful in some respects, in others it was not. In his extensive study of the technology transfer between Hungary and Western countries Paul Marer shows that despite its comparatively wide scope the NEM was not able to overcome the technology gap between Hungary and the OECD countries. Between 1968 and 1983 the ratio of Hungarian high-tech export products to those with a low technology content roughly stayed the same. The absorption and dissemination of technology was slow, the Hungarian export successes in Western markets limited.[22]

The inability of the NEM to overcome the structures of central planning hurt the S&T system badly. In the early 1980s a Hungarian research team analyzed the industrial R&D sphere of the country and came to the conclusion, that

> (a)ccording to our findings, the enterprises are not sufficiently interested in the economic success of the research and development actions. The analysis of expectable efficiency hardly motivated their decisions taken before and during the realization of the innovation. The improvement of the economic efficiency of research and development actions cannot be really expected from compulsorily calculated indicators. (Inzelt 1982, p. 51)

Some of the underlying reasons for this partial failure of the NEM in respect to S&T seem to be rather obvious, whereas others do not seem to be so clear. One of the major problems of the industrial R&D system in centrally planned economies that was not effectively overcome was the isolation of R&D from production. A key argument in explaining the failure to overcome this compartmentalization was that, despite a few closures of companies over the 1980s, all economic units in Hungary still faced soft budget constraints until the transition to the market economic system began in 1989.

Due to the reshuffling of costs and benefits by the state in the centrally planned economy, companies only faced soft budget constraints; they did not have to fear bankruptcy and at times they still were judged by output measurements, typical for Soviet style economies, rather than by standards of efficiency or profitability, typical for capitalist economies. Although profitability was a central concept to the NEM,[23] it was hard to actually measure it in a system still characterized by the redistribution of costs and benefits throughout the whole economy.

Companies did not feel compelled to react to consumers' needs because of the soft budget constraints. They were still acting in the environment of the seller's market of a shortage economy, where demand for goods is in general is lower than production capacity. In such a market buyers never gain the possibility of choosing between goods, because their basic demand is not fully met, i.e. the consumers buy what they can get.[24] Consequently, the managers had more reason to worry about their output figures and about pleasing their superiors than about consumer demand (Keren 1992, p. 107).

The controlling of companies was administered by the public bureaucracy. But as these authorities had no instrument at hand to measure the efficiency of the companies in question, these companies had not to face an effective control system. Moreover, it is not clear if the public bureaucracies themselves were interested in an innovative behavior of industry. Under the constraints of the output measures mentioned, above bureaucracies could maximize their profits better by extensive factor usage better than by intensive factor usage. In other words it was more suitable to raise the quantity of input, namely capital, land, labor and technology, than to raise the quality of input, as for example through innovations.

Such a strategy of simply enlarging the inputs of the economic system in order to achieve short term growth[25] explains the large debt load acquired by Hungary over the 1970s and 1980s. The Hungarian debt figures of the time of the NEM can be interpreted largely as an effect of politicians, bureaucrats and managers who maximized their behavior individually in the fashion described above and a system which did not allow for a reasonable amount of checks and balances.

The basic difference between a classic centrally planned economy and a reform socialist system as they have been described here is that in the reform socialist system elements of a market economy are allowed to have a controlled influence on the economic system. One of these elements may be the bureaucratically simulated market, as in the first phase of the NEM. Another element may be that of privately owned enterprises, mostly belonging to the small and middle enterprise category, as was the case in the latter phases of the NEM. In neither the unreformed nor reformed centrally planned economies can one find a market mechanism effective enough to be stronger than the bureaucratic hierarchies or informal networks.[26]

While the central role of bureaucratic hierarchies seems evident, it may be somewhat surprising that networks play a strong role in centrally planned economies. Nonetheless, Katalin Balázs points out that in order to get an innovation implemented in industry, 'researchers often used their close personal and institutional contacts with the authorities to solicit administrative pressure on the firms' (Balázs 1993, p. 543).

Janos Kornai has analyzed the control which bureaucratic hierarchies had over the economy. He insists that neither the 'direct bureaucratic control' of the classic Soviet economic system, nor the 'indirect bureaucratic control' of the reform-socialist economic system as featured by Hungary after the introduction of the NEM, were beneficial for the respective national economy. Kornai places market economies and reformed centrally planned economies on a continuum between market and bureaucracy controlled systems. In each type of economy can one find elements of the other. 'The real issue is the relative strength of the components of the mixture' (Kornai 1990, p.117). At a certain point the bureaucratic controls are stronger than the market and the firm's managers are influenced more strongly by higher authorities than by the market.

A number of indications can be found for the hypothesis that even during the heydays of the NEM, Hungary was definitely closer to a centrally planned than to a market economic system. In 1986 Paul Marer analyzed the system of economic regulators, the institutional structure of the economy and the decision-making process for production, trade and investments, and came to the conclusion that while Hungary was not longer a 'traditional', Soviet style, centrally planned economy anymore, it still had a long way to go before it could become a market system. Marer gives an account for the amount of the planner's interference with the economy, which becomes evident most clearly in the cases of mismanagement as follows:

- huge investment projects were begun simultaneously, overextending the country's resources by far,
- exports into the COMECON area and the meeting of domestic demand were prioritized, but as an effect of the severe deterioration of the terms of trade with the Soviet Union during the 1970s little indigenous investment was left for the production of exports into the capitalist countries with their important convertible currencies,

– too few efforts were made to increase the technology content of production, too much emphasis laid on the output of basic materials and semifinished products, and, finally,

– the level of investment had to be cut by the government, when the size of foreign debts had grown to an almost unmanageable extent.[27]

Another indication for the degree of the bureaucracy's meddling with the economic system is that besides the general planning of the economy, Hungarian authorities still interfered in the economic day-to-day business during the NEM (Chilosi 1992, p.178).

The organization of industry until the latest transition of the economy is yet another indicator for the inability of the NEM to overcome the main deficiencies of the central planning is. Hungary's industrial structure was significantly overcentralized in comparison to both the other COMECON and the capitalist Western countries. One might be led to believe that the planners of the NEM would have been particularly interested in the decentralization of the firms or the creation of a small and medium sized enterprise (SME) sector during the first years of the reform. After all, the NEM took so many other elements from the market system. Surprisingly, the contrary was the case: the concentration process of the Hungarian economic system was still going ahead in the 1970s, as can be seen in table 5.7.

Table 5.7 **Concentration of Industrial Organization in Hungary, 1950–1980**

Year	State Enterprises		Cooperatives		Private Artisans (thousands)
	Number	Average Employed	Number	Average Employed	
1950	1337	337	n.a.	33	n.a.
1960	7386	639	1251	110	82
1970	812	1374	821	226	61
1980	699	1569	661	287	46

Source: Marer 1986, p.36.

The exceedingly high concentration of production capacities in Hungary's economy changed only slowly in the early 1980s. From 1981 on large company holdings were dismantled in Hungary. This move to decentralize the firm structure at the top end was complemented by the liberalization of the SME sector, once the creation of private firms and cooperatives had been allowed by the authorities (Inzelt 1986, p.179).

But not only the organization of firms at the macro-level was largely unchanged by the NEM. The micro-level organization of the enterprises after a decade of reforms in the late 1970s was not decisively different from the organization of firms in the late 1950s. The organization of production was still centralized. A number of functions, which in capitalist systems were performed by the firms themselves, as for instance strategic planning, were still delegated to the authorities. Finally, functions such as R&D or commercial activities lay in the hands of seperate institutions (Inzelt 1986, pp.184). Despite repeated efforts in all these areas to make substantial changes, not one issue was ever completely overcome until Hungary's last transition to capitalism had begun.

It is hard to say what would have happened if the New Economic Mechanism (NEM) in the early 1970s had not been restricted again, but had been able to extend its scope. It is, however, safe to say that the chances for such a process were not exceedingly high. In fact, the discussions over the reforms sometimes led to bizarre results. The control over the strategic planning of the enterprise sector is a good example: In 1973 the Economic Commission was dissolved and replaced by the State Planning Commission, which had a different set of tasks, but which was also established to guide the economic development of industry. In 1979, after the reinvigoration of the NEM, the Economic Commission was revived, and was now to work side-by-side with the State Planning Commission.[28]

The constant struggle over economic competencies and reforms found its expression in the S&T sphere. Apart from the insecurities which are always to be found in a situation where structures and functions of a system are regularly, if not fundamentally, changed, researchers found themselves on two sides: one favoring the reforms, the other opposing any further changes. One example for this can be found in an article of the then well-established late social scientist Péter Vas-Zoltán, who wrote,

(T)he problem that scientific and technological progress may possibly have certain negative side effects that act against societal objectives has not been dealt with in depth so far. To quote an example, 'innovation' is now regarded in Hungary as having a significant 'value', but little thought is given to the social inequalities that would result from a more effective and rational national economy. (Vas-Zoltán 1985, p. 286)

To sum up, one can say that the Hungarian NEM had reached its goals only insufficiently. Ferenc Jánossy's afore-mentioned criticism on the country's economic system, written in 1969, would have still been correct fifteen years later (Jánossy 1969). The country was 'quasi-developed'. Worse still, despite all the reform efforts, Hungary of the mid-1980s had not gained on its capitalist Western European neighbors. For the S&T system this meant a clear deterioration of its infrastructure. Since the number of researchers had increased over the 1970s, while total spending in most sectors hardly had grown, the percentage of investments had to be reduced substantially, from 21% in 1971 to 12% in 1981. A result of this situation was that it was possible to print good looking figures on the statistical reports, but the reality behind these reports was Jánossy's 'quasi-development'.

Under the load of such technological and economic backwardness, the intrasystemic resistance to more radical reforms diminished during the 1980s. The latest transition of Hungary, this time back to capitalism, could begin.

Notes

1   Haslinger (1996, p.261) cites a report of the embassy from June 4th, 1968.
2   While this may seem reminiscent of problems which the US National Endowment for the Arts (NEA) has persistently featured over the years of its existence, there are differences in the scope of both problems. Whereas the NEA in the 1980s and 1990s had to face oppositition by conservatives like Senator Jesse Helmes in matters such as the funding of homosexual or lesbian artists, an interview partner working for a central institution of the Hungarian S&T system pointed out that until as late as the 1970s she was confronted with an intervention from the Central Committee of the MSZMP because she wanted to quote figures from the Hungarian budget in an international publication ('our enemies in the West could use the numbers against us...'). For an account of the discussion in the US about the funding of art see Zambra, Karin, 'Wechselwirkung zwischen der Performance Art und der Politik in den USA' (The Relationship between Performance Art and Politics in the United States), Mag. phil. thesis, University of Vienna, 1994.

3   Compare with the open criticism of István Bessenyei in Bessenyei/Melchior 1996, pp.229.

4   There was a general consensus on this view among all interview partners confronted with this question, of whom many were knowledgeable about the situation in other realsocialist countries as they had travelled extensively in the Eastern Bloc as researchers.

5   Az MSZMP Tudománypolitikai Irányelvei (Science Policy Guidelines of the Central Committee of the Hungarian Socialist Worker's Party), Kossuth Könyvkiadó, 1969, cited in: Peter Vas-Zoltán 1985, p.283.

6   As a number of interview partners have pointed out, with the exception of a few taboos such as the presence of the Red Army in Hungary, research topics could be chosen freely by the end of the 1960s. Despite this development, during the roll-back of the NEM's reforms the freedom of expression was also limited again. For instance, a number of social scientists who at the beginning of the 1970s attempted to reinterpret scientific Marxism were subsequently excluded from Party membership and faced with disciplinary measures.

7   It is interesting to see the parallels between the development of ideas concerning the role of the state in R&D in Western and Eastern Europe. While in the West the Rothschild report in the UK led to less en bloc funding and more individual contracting out of R&D institutions, the same can be said about the Hungarian NEM of 1968 – only that the Hungarian plan went into effect several years earlier.

8   Marer 1986, pp.220; however, one has to keep in mind that there are sectors that fared better than others; Inzelt 1994 describes successful sectors (pharmaceuticals is an example); another example which has been pointed out to me by several interview partners and which is mentioned by Marer, too, is agriculture. Negative examples are the information-technology-related sectors.

9   For the reasoning behind a closer cooperation between Hungary and Austria, see Haslinger 1996, pp.255.

10   See the discussion of the 'third path' between capitalism and communism in the introduction and the chapter 'The Hungarian Reform Process: Visions, Hopes and Reality', in: Kornai 1990.

11   For a discussion of the original intentions of the policy makers and what reality looked like in the Hungarian post-NEM S&T system of the 1980s, see Balázs 1993.

12   See Marer 1986, p.15/16. As was pointed out to me repeatedly by persons from former Eastern Bloc countries like Russia or Uzbekistan, Hungarian products were known in the COMECON countries for their comparatively good quality. In addition to this, Hungary's tourism industry flourished not

only because of the wonderful sights and landscapes of the country, but also because for citizens from most other Eastern Bloc countries Hungary came very close to a Western market oriented country in terms of culture, life-style and availability of goods.

13   It is especially interesting that the rising R&D expenditures were criticized by a researcher working for the MTA, Mária Csöndes; see Csöndes 1985, p.184, and, again, in Darvas 1988, p.151.

14   After the first oil crisis in 1973 (and the Vietnam War), the two super-powers were neither willing, nor capable of internalizing the costs of the Cold War anymore. This was the beginning of the realignment of the powers and the rise of a multipolar world order.

15   In interviews I was given figures between 25 and 50% for the amount of funding basic science oriented MTA institutes had to acquire through contracting out of R&D from other organizations including industry.

16   Stories about the economic activities Eastern European S&T institutes have had to engage in since the mid-80s are legion by now. There have, for instance, been reports about scientists originally doing research in electrical engineering who are now selling satellite dishes to make a living.

17   Balázs 1992, p.91.

18   It should be noted that growth and quality of scientific work are not necessarily found side by side. Some areas of scientific research in Hungary are known for their excellency, whereas others are in still early stages of their development when compared to most OECD countries.

19   Precisely this uneven development of the different disciplines shows the limitations of the 'demand-pull' notion. Research areas promising results in a rather short time frame get funded by industry, whereas basic research and the social sciences are – without additional funding possibilities – left out in the cold.

20   One interview partner pointed out that the cooperation between the two institutions is remarkable for two reasons: one is the division of labor, with the university performing basic and applied research and MEFI engaging into applied research and, primarily, development and engineering. The second reason is the length of the relationship that has lasted now for almost four decades. It has led to a research group at the university linking basic and applied science.

21   Darvas claims 1985 (Darvas 1988, p.151), Balázs et al claim 1984 as founding date for the OTKA (Balázs et al 1990).

22   Marer 1986, see especially the summary of his findings on pp.14.

23   See Rezsö Nyers, one of the leading personalities of the NEM, quoted in Kornai 1990, pp. 60.

24   For a characterization of the problems of a shortage economy, see the classic work of Kornai (1980).

25   As conservative critics will ascertain, such a strategy is not uncommon to Western Socialdemocratic governments either, as for example the Austrian government during the 1970s, whose political system was dominated by the 'Austrokeynesian' Kreisky administrations which amassed a sizeable national debt. Of course, as Socialdemocratic critics will ascertain, such a strategy is not uncommon to Conservative governments either, as for example the US government during the 1980s, whose political system was dominated by the 'Neoconservative' Reagan administrations, which were supposedly Monetarist, but in fact amassed a sizeable national debt, too. For an analysis of the economic strategies of right- and left-wing governments and their effects on the national debt, see Wagschal 1996.

26   For recent comparisons of the different forms of social order see for example, Streek/Schmitter 1985; Hollingsworth/Schmitter/Streek 1993; Grabher 1993; in the volume which Grabher edited the Hungarian economy serves as an example for a system with a strong network component, see Grabher 1993, p.20, and Neumann 1993, pp.179.

27   Marer 1986, for the general statement of affairs see the executive summary on pp.14, from which this list was drawn; for an example on how the interventions of the public bureaucracy changed over time, see pp.38.

28   Further information on the frequent changes in economic policy making in Inzelt 1986, pp.185.

6 The Third Transition: The Return to Capitalism

6.1 Historical Development of the S&T System

The latest transition of Hungary back to capitalism may have begun in 1988/89 with the gradual realization of the Communist Party's leadership that its time had come to open the way for a system change. The transition may have begun earlier, in 1988, when Janos Kádár, the long-time General Secretary of the Communist Party and leader of the state, was forced to step down and clear the way for a new generation of politicians. The transition may have begun even earlier, in the beginning of the 1980s, with the changes in the legal and economic system of the country which made international joint ventures and other forms of foreign influence possible. For the purposes of this analysis the beginning of the latest transition Hungary went through, from reform socialism back to capitalism, will be assumed to have started in spring 1990, when the first free democratic elections after more than four decades were held.

The last decade of the People's Republic was marked by stagnation slowly changing into slight recession. After the downfall of realsocialism in the years of 1988 and 1989, the economy faced the same shock that all formerly realsocialist economies had to cope with when they opened up their systems. The economy first began to shrink slowly, and then went downhill faster than most observers – and the people themselves – would have expected. Looking back, the main reason for the surprisingly steep recession was the faltering market for many products in Hungary. Due to the switch from a seller's to a buyer's market, companies producing unwanted goods soon had to stop production and finally were closed. Consequently, unemployment went up steeply. The decline in their production, together with the shrinking GDP figures reflected in table 6.1, was steeper than the one of the Great Depression (Kornai 1993, p.2).

103

Table 6.1   Indices of GDP and Industrial Production in Hungary, 1986–1997

Year	GDP	Mining, Manufacturing, Electricity
		(Previous year = 100)
1986	101.5	99.7
1987	104.5	103.2
1988	99.9	98.5
1989	100.7	98.0
1990	96.5	92.4
1991	88.1	82.2
1992	96.9	93.3
1993	99.4	103.0
1994	102.9	106.0
1995	101.5	107.0
1996	101.3	103.5
1997	104.4	111.1
1998	105.1	112.8

Source: Hungarian Central Statistical Office 1997, p.245; numbers for 1997 and 1998 are from the CSO's WWW-site with the address http://www.ksu.hu/eng/free/ and were retrieved on 16–02–1999.

Note: From 1992 on the classification of 'Mining, Manufacturing and Electricity' has been changed and is in accordance with the Hungarian Standard Industrial Classification (equaling ISIC Rev. 3 at 2-digit level).

The underlying reasons for the economic slump include the change from one socio-economic form of organization to a completely different one with subsequent changes in the demands of society with respect to the qualities of products, skills of people and the like. Society and economy had to reorient themselves to a new mode of production. Previously, capital and consumer goods had been produced with machinery and other capital goods imported from the West and were then exported to the East. Due to the political and economic collapse of the COMECON, the Council for Mutual Economic Assistance, and the Soviet Union, exports had to be rearranged, as table 6.2 demonstrates. In only four years, from 1987 to

1991, the ratio of exports to the European Union (EU) as compared to the exports to the COMECON had been reversed: whereas in 1987 20.1% of total Hungarian exports went to the EU, in 1991 this figure had more than doubled to 45.7%. Correspondingly, the export figures to the COMECON area shrank to less than half of its original size, from 45.9% in 1987 to 19.4% in 1991.

Table 6.2   Destination of Hungary's Exports, 1987–1993

Year	European Union	(Former) COMECON Area
	Share of Total Exports (%)	
1987	20.1	45.9
1988	22.5	48.2
1989	24.8	35.7
1990	32.2	36.0
1991	45.7	19.4
1992	49.7	22.9
1993	46.5	22.8
1994	50.9	22.1
1995	62.8	23.3
1996	62.8	23.5

Source: Central Statistical Office 1997, p.279; own calculation.

In the first half of 1996, 59% of Hungary's exports went to its five largest trading partners, the largest being Germany, followed by Russia, Austria, Italy and France. Around 56% of imports came from these countries. Total exports consisted of 49% processed goods, 26% machinery, 21% foodstuff and tobacco and 4% energy. Imports were made up of 49% processed goods, 29% machinery, 12% energy, 5% foodstuffs and 5% raw materials.[1] By way of comparison, in 1997 64% of Hungary's exports went to its five largest trading partners, the largest again being Germany, followed by Austria, Italy, Russia and France. Around 58% of imports came from these countries. Total exports consisted of 45% machinery, 36% processed goods, 13% foodstuff and tobacco, 4% raw materials and 3% energy. Imports were made up of 42% machinery, 41%

processed goods, 10% energy, 4% foodstuff and tobacco and 3% raw materials.[2]

These figures clearly demonstrate that Hungary features a developed industrialized national economy with certain deficiencies in high and medium technology sectors. They also reflect the growing importance of the EU as Hungary's trade partner and the rapidly re-emerging machinery sector as an export leader – which is mainly foreign direct investment (FDI) driven. The observed changes are also a reaction to the demand structures in the regions newly targeted for exports. For the industrial R&D system all these changes in macro demand and production structures meant the necessity to react to the need for goods produced under higher norms and standards.

The economic changes Hungary was going through, were contingent not only on economic, but also on political variables. In the spring of 1990, the first free elections after more than four decades brought a coalition of conservative parties, headed by Prime Ministers Antall and, later, Boross, to power that transformed the economy relatively cautiously. Over the next years the Hungarian Parliament came to feature three committees of relevance to the S&T system. These were the Economic Committee, the Science and Innovation Subcommittee and the European Integration Committee, of importance because of the Hungarian EU accession efforts.

A course of relative budget austerity was followed, the newly created State Property Agency began its sales of the nationalized companies and a stock market was opened. Property rights were established, laws on bankruptcy and accountancy procedures, patents, higher education and the Academy of Sciences, the MTA, were passed. A number of these new laws and decrees, especially the laws on Higher Education and the Academy of Sciences from 1994 had quite a direct influence on the S&T system – mostly along the lines of the recommendations of the OECD evaluation of 1992 (Biegelbauer 1994). One example is the dialogue between the MTA and the universities, which was an outcome of the discussions on the new laws, which divided the turf of both organizations anew. The unconventional two-level system of scientific degrees[3] is not in effect anymore. In the summer of 1993 the granting of the Ph.D. was finally referred back to the universities.[4]

However, despite such efforts a real reorganization of the system – other than its downsizing – has been only gradually taking place. To make things worse, little money has been spent on the S&T system by government and private industry likewise. Table 6.3 shows a steady decrease of R&D expenditures from 1990 to 1994.

Table 6.3   National Expenditure on R&D as Percentage of GDP, 1990–1996

Year	1990	1991	1992	1993	1994	1995	1996
National R&D Expenditure/GDP	1.6	1.1	1.1	1.0	0.9	0.8	0.7

Source: Central Statistical Office 1997, p.471 and own calculation; OMFB 1998, p.5.

According to a number of experts, the R&D expenditure is expected to stabilize itself between 0.6 and 0.7% of GDP for the rest of the decade. This is a truly low level, not only in lieu of the governmental S&T policy documents, but also in comparison to other highly industrialized countries.

The next table shows how the internal structure of the R&D expenditures has changed. In a time of rapidly shrinking S&T budgets, the state has withdrawn from this policy field even faster than has the industry.

Table 6.4   R&D Expenditures by Source as Percentage of GDP, 1990–1996

Year	1990	1991	1992	1993	1994	1995	1996
Business	0.62	0.44	0.34	0.29	0.27	0.28	0.26
State	0.93	0.61	0.68	0.65	0.59	0.44	0.36
Other	0.05	0.04	0.06	0.08	0.07	0.04	0.05

Source: Central Statistical Office 1997, p.471; OMFB 1998, p.5.

As holds true for other formerly centrally planned realsocialist systems, too, the number of scientists and engineers in company R&D units has fallen dramatically.[5] This development is clearly linked to the reduction of the R&D expenditures by both business and government. When comparing the numbers of scientists and engineers working in companies with those of other R&D units, it becomes obvious that industrial R&D has suffered much higher losses than has R&D in other sectors.

Table 6.5   Number of Business Enterprise Researchers and Numbers Expressed in Percentage of National Total Number of Researchers, 1990–1995

	1990	1991	1992	1993	1994	1995
Company	7.629	5.341	3.724	3.503	3.330	2.926
Percentage of National Total	43.5	36.9	30.2	29.6	28.3	27.9

Source: OECD 1997, pp. 28.

To hint at what has happened to the national R&D output, table 6.6 looks at the total numbers of patents registered in Hungary and abroad.[6]

Table 6.6   Number of Patents Taken Out in Hungary and Abroad, 1990–1993

	1990	1991	1992	1993	1994	1995	1996
Hungary	641	576	422	272	183	190	188
Abroad	838	556	539	467	475	412	273
Total	1479	1132	961	739	658	602	461

Source: Central Statistical Office 1997, p.467.

Other output indicators of the national R&D effort will be discussed in the next chapter. However, it is important to keep in mind that the methods of statistical measurements are sometimes problematic in Hungary. A variety of problems make the statistics – especially in relation to industrial R&D – unreliable. To begin with, the statistical data used during the time of the People's Republic were overestimating the total amount of R&D in comparison to the OECD data as laid down in the Frascati and Oslo Manuals. Worse, however, is that among the data compiled two changes in standards have occurred since the beginning of transition: first from the

COMECON system to the UNESCO system, then to the OECD system. A bit more light shall be shed on the problems of Hungarian transition statistics in the discussion of the financing of R&D in industry later on in this chapter.

6.2 Policy Maker's Models and Intentions

Although the means of the center-right Antall/Boross administrations and the center-left Horn administration often differed, the governments' goals for both periods were identical: transforming society and economy into a capitalist system and catching up with Western Europe. Another thing both time spans had in common was that despite the lipservice both administrations paid in a number of documents and speeches, saying that the S&T system was crucial to the success of the national economy, it was not treated as a preferential policy area.

In an analysis of the policies implemented since the Hungarian elections in 1990 three phases can be distinguished.[7] 1990 and the first months of 1991 were marked by the same euphoria that flooded Europe, East and West, immediately after the fall of the iron curtain. The expectations of policy makers and the general public was that the market forces, once unleashed, would generate an order of their own.[8]

During this 'phase of euphoria' the Hungarian government reacted to the problems of the S&T system primarily by initiating changes in the structure of the system. As was the case in other Central and Eastern European Countries (CEECs), too, the S&T system was granted representation on the highest levels of government: two new ministers were established, one for science, one for technological development.[9] The largely uninfluential 'Science Policy Council' of the Antall/Boross governments, replacing the 'Science Policy Committee' (TBP) in 1990, was headed by a chairman who was at the same time Minister of Science without portfolio. The President of the MTA was still represented in the Science Policy Council, but the MTA's General Secretary not any more. Besides the Science Minister and the President of the MTA, the Minister of Culture and Education, the President of the OMFB – at the same time Minister for Technological Development, and the Minister of Finance had a seat in the Council.

In stark contrast to the new representation of S&T at the highest levels of government stood the funding for the S&T system, which decreased.

Again, the government expected the invisible hand of the market powers to bring about a balance of supply and demand for R&D. The 'demand-pull' model's influence still was strong in the minds of the policy makers.

The next phase was characterized by the realization that the obstacles on the road to a market economy were bigger than expected. In this 'phase of frustration', which took place from the second half of 1991 to the first half of 1993, the decline in levels of production which was due to the political upheaval in the area of the (former) Soviet Union, the subsequent loss of markets, the rising unemployment levels and various interlocked economic, social and political problems led to a widespread frustration about the prospects of the immediate future, not only in Hungary, but in all CEECs. The fact that living conditions were deteriorating in large parts of society led to a comeback in power of some of the political offsprings of the former Communist parties in the region in the second half of this phase. These parties were regularly confronted with Cold War-thinking from Western Europe and the United States.[10] Contrary to the fears of Western observers, it has turned out that most socialist parties in transition countries have changed from realsocialist planners to neoliberal pragmatists as soon as they have come into office.[11]

The S&T policies of the Hungarian government during the 'phase of frustration' changed only marginally. During this time crisis management was the main task of the authorities. Because of the weakness of industry and its inability to fund R&D, the government was forced to step in and increase its support for industrial R&D. The National Committee for Technological Development (OMFB), in particular, began to implement a policy of allocating the Central Technological Development Fund's (KMÜFA) capital primarily to near-market innovations. The diminishing of total state expenditures on R&D has been slowing down since 1992, but a rise in expenditures is still not in sight.

The third phase may be called the 'phase of realism'. Since the second half of 1993 on, East and West European[12] governments and the general public alike have learned from the experience made in the transition period. Expectations in most cases are not unrealistically optimistic or pessimistic anymore. Medium and long term strategies of governments now have been formed.

During the first years of the transition from realsocialism to capitalism the Hungarian S&T organizations all were subject to a huge pressure to reorient their structures and functions according to the needs of the new capitalist system. The small National Committee for Technological Development (OMFB), with a staff of 170, is a good example for an organization reacting quite flexibly to the changing environment. During the transition period OMFB had a steep learning curve. Beginning with very modestly structured programs in 1990, the organization has applied a competitive funding scheme primarily for industrial R&D since 1991. In addition to this, the organization has regularly used outside expertise, as for example an evaluation of the applied R&D system by the Swedish NUTEK institute and the Academy of Sciences (MTA) Research Institute on Industry and Enterprise Economics during 1995 and 1996, for the restructuring and evaluation of its programs. Moreover, the official publications of OMFB mirror the increasing 'realism' of the organization. Whereas, in a text from the institution's phase of euphoria, in 1991, buzz-words such as 'artificial intelligence', 'biotechnological processes' and 'robotics' were used for the description of the organization's priorities, in 1994 a similar text more modestly uses 'infrastructure development' (in 1991 the last item in the list, in 1994 the first), 'agriculture and food processing' or 'promotion of small and medium sized enterprises', which is a reflection of the organization's entering into the 'phase of realism' (OMFB 1991, p.4 and OMFB 1994, p.6).

During this 'phase of realism' government publications and statements of policy makers began to acknowledge the necessity to intervene in cases of market failure.[13] The Hungarian policy makers have switched from the linear to the complex holistic model of technological change, resulting in a pragmatic mixture of policies now taken in eclectic manner from other countries. The codification of the S&T system was making progress due to laws and decrees aimed at higher education, the Academy of Science (MTA), the National Fund for Scientific Research (OTKA) and the National Committee for Technological Development (OMFB). Moreover, the government issued innovation policy guidelines on May, 7th 1993 (OMFB/IKM/Ministry of Finance 1993). After a long discussion process full of conflict, a long-term technology concept was worked out by OMFB in 1995, followed by a government action plan in early 1996. Despite its criticism, the allocation of projects has become increasingly more competitive and open. Policy goals for issues of S&T have been primarily of short and medium range. The main emphasis, especially of the funds allocated through KMÜFA, until 1995 has been on infrastructure programs.

Some funds have been used to counter the brain drain and are specifically aimed at young researchers.

From a structural point of view, the signals which the Hungarian government sent to the S&T system were mixed. On the one hand, neither science nor technology were represented at the ministerial level during the Horn government anymore. On the other hand the 'Science Policy Collegium' was founded in 1994, an advisory body to the newly elected Prime Minister Horn, who also chaired the organ. The Collegium had three new members besides the Prime Minister, namely the presidents of OTKA, of the Rector's Conference and of the College Director's Conference. Similar to the fate of its predecessors, the body had only a limited impact on the S&T system. But, at least, the Science Policy Collegium met regularly, which had not been the case for its predecessor. Several experts have pointed out that the Collegium was dominated by the MTA, which in terms of budgetary criteria fared relatively well during the second half of the Horn government.

The one part of the S&T system which did not fit into the time frame of the model laid out above, featuring three distinct phases of development during the first years of the latest transition to capitalism, is the higher education sector. As will be shown later, the universities in particular, fared better than both industrial and the MTA-bound R&D – specifically until 1995. In 1995 the austerity program of Finance Minister Lajos Bokros called for sizeable cuts in the national budget, and for the first time since 1990 these cuts included universities. While the universities had already been through their 'phases of euphoria, frustration and realism', by mid-1995 they had fallen back into a phase of frustration.[14] Their surprise over the reduction of university personnel by approximately 15% was great.[15]

Still, the universities have been better off than the other sectors of the S&T system. From the beginning of the latest transition on they were given more resources and received more interest from the side of policy makers. As early as 1989, even still before the magic date of the fall of the Berlin Wall, Marxism-Leninism in its previous form had been eradicated from the curricula of the schools and universities. Learning the Russian language was not compulsory any more.[16] The Soviet model of the higher education system, insofar as it had not been reformed yet, was finally discarded now.

The restructuring of the higher education sector was begun with the model of the German university system in mind. The complete unification of teaching and research on the one hand and the condensing of research potential in large universities on the other hand was envisioned by policy makers (Melchior/Bessenyei 1996, p.49). However, the process of reorganization of higher education units with the goal of creating universities, colleges and 'Höhere Technische Lehrschulen'[17] in the Western sense is still in progress. Therefore, the massive fragmentation of the higher education system – there still are institutions which include only one department or faculty – will not be solved in the next years.

Policy makers were frequently looking to foreign countries as examples for the building of structures and functions of the S&T system. The applied science oriented Bay Zoltán institutes have been formed after the German Fraunhofer Gesellschaft institutes. For the rationalization of peer-review and project evaluation procedures institutions such as the Deutsche Forschungsgemeinschaft (DFG) or the United States' National Science Foundation (NSF) have been models.

6.3 Actual Effects of Policies

The National Committee for Technological Development (OMFB) in the very beginning of transition expanded in terms of personnel, leverage and budget. It was elevated to cabinet level during the conservative Antall/Boross governments. Because of the repeated budget cuts of the OMFB's main instrument, the Central Technological Development Fund (KMÜFA), this rise in size and influence was only relative. A few years later, in 1994, the left-of-center coalition under prime minister Horn, formed by the socialist party (MSZP, one of the offsprings of the Communist Party, the MSZMP) and the Alliance of Free Democrats (a center to center-right party, SZDS), reorganized the national governance structures again.

After being scaled back from a ministerial to a lower level, OMFB became an independent agency headed by a president and supervised by a council with a chairman. After a struggle which lasted several months OMFB is now supervised by the Minister of Industry and Trade (IKM) without being an actual part of the ministry. OMFB is still responsible for the elaboration and coordination of national innovation policies (Inzelt 1995a, pp. 30). However, the small agency has had to hand over the control over the Office of Standardization, the National Meteorology Office and the National Patent Office to the Ministry of Industry and Trade.

OMFB is essential for the funding of industrial R&D. For this purpose the organization has been utilizing the Centralized Technical Development Fund/Program (KMÜFA), which has been cut back several times rather drastically since the beginning of the latest transition. In the five years between 1990 and 1995 KMÜFA sank even in nominal terms. Taking the high inflation rates for this time period into account, the fund was reduced by approximately half its volume. The low 1996 and high 1997 figures, reflected in the following table, are an effect of budgetary operations of the Horn government, which held back part of the KMÜFA finance in 1996, but decided to disperse these funds in 1997. In any case, even with the reduced funds KMÜFA still is the single most important source for industrial R&D in Hungary.

Table 6.7   Allocation of KMÜFA Funds in 1990-1995, Nominal Funds
            (HUF million)

1990	1991	1992	1993	1994	1995	1996	1997
3.75	4.45	6.88	8.08	6.46	4.12	3.35	9.83

Source: http://www.omfb.hu/web/ retrieved at 16–02–1999.

The five main programs under which the allocation of KMÜFA funds has taken place are listed in table 6.8. Regardless of the answer to the question whether the funding scheme of OMFB or the proposal of the OECD in 1992/93 were first developed, it is obvious that the recommendations produced by the OECD regarding the allocation of funds and the actual allocation of KMÜFA funds are in most parts similar (Biegelbauer 1994). This holds especially true for OMFB's infrastructure improvement program, a central proposition of the OECD team. However, it was specifically this program that suffered most from the recent budget cuts of the Horn government.

Table 6.8   **Allocation of KMÜFA Funds in 1993, 1994 and 1995** (HUF million)

Program Names	Funding Size 1993	Funding Size 1994	Funding Size 1995
R&D Infrastructure Improvement Program	3700	2640	99
Applied R&D Program	2087	2000	1907
National Projects Program	1185	1566	410
Export Promotion Program	120	343	198
Patent Licensing Program	25	32	64

Source: OMFB 1994, p.8; OMFB 1995, p.8; OMFB 1996, pp.8.

The granting of support for the applied R&D program depends on the willingness of the applicants to invest their own funds, which is a prerequisite for funding by OMFB. The national projects are diverse in nature, among them are geographic information systems, the depositing of low and medium radioactive nuclear waste and food processing technologies. These projects are jointly administered by OMFB and other authorities on the national level. The export promotion program which aims at export-oriented product development is coadministered with the Ministry of Trade and Industry. Finally, the patent licensing program supports the licensing of Hungarian patent applications abroad.

Two other funding activities of the OMFB are rather small, too. The first is the sponsorship program for improving the social conditions of RTD, which mainly aims at covering membership fees of applicants at international organizations, traveling costs and other international cooperation oriented expenses. The second is a separate competition system supporting the participation of Hungarian R&D units in the Framework Programmes of the European Union. Furthermore, in 1997 two funding initiatives have been added to the tasks of the OMFB: a competition program for new information and communication technologies and a tender program for regional innovation activities.

The application review process for these programs has been mastered according to the peer review system, with a board of experts making the final decision about the support of a project. OMFB points out that the funding of projects may either take the form of nonreimbursable support or preferential loans or a mixture of both, 'depending on the degree of risk involved' (OMFB 1994, p.8). Most funding through KMÜFA takes the form of loans. It is not entirely clear how 'soft' the character of the KMÜFA loans for businesses actually is.[18]

Shortly after being voted into office, in the summer of 1994, the Horn administration aired a plan to pool all 32 S&T and economic funds – including OMFB's Centralized Technical Development Fund (KMÜFA). The reasons given for the reform were that the large number of funds existing in Hungary were mismanaged during the Antall/Boross governments and could not be supervised well enough by the authorities. Moreover, a rationalization would be adequate to the scarcity of money in the governmental budget. This governmental plan has caused a great deal of discussion throughout the S&T system. After all, since the demise of the MÜFAs the KMÜFA controlled by OMFB has been the most substantial centrally coordinated fund for R&D purposes in Hungary.[19]

As a result of the internal quarreling of interest groups, the 'Economic Development Fund' (GFA, Gazdaságfejlesztési Alap) was not created until 1996. KMÜFA, now the 'Basic' Central Technical Development 'Program', has become a budget title of the large new fund. Since then, OMFB has lost control over the size of KMÜFA, which is subject to decisions first of the Council of Ministers and then the Minister for Trade and Industry, respectively.

The main concern of people working in the S&T system regarding the reorganization of the KMÜFA was that the representation of their interests had grown so weak during the first years of the transition period that they would not be able to get a fair share out of a single fund from which many more powerful interest groups would also be financed. In fact, political interest representation of the S&T system's institutions grew more difficult after the end of realsocialism in Hungary. After all, science and technology were given priority by the realsocialist governments even in a situation of economic stagnation. Moreover, the structures of government were favorable for the S&T system. The Academy of Sciences (MTA) was de-facto a science ministry. The National Committee for Technological Development (OMFB) was an independent agency, entrusted with the single largest S&T fund of the country. The rivalries between the MTA and the universities as well as the MTA and the OMFB had no negative effect on the interest representation of the S&T system during the 1960s and 1970s.

This situation changed dramatically when the support from government and industry began to decrease decisively during the second half of the 1980s and the beginning of the 1990s. All of a sudden capital was scarce for S&T. The rivalries between the different institutions of the S&T system intensified. The MTA was accused of being large and wasteful, of having too many functions and of being uncooperative especially towards the universities (Darvas 1995, pp.63). In 1988/89 the first free trade unions of the People's Republic were formed, representing the interests of the personnel of the MTA. Since about this time the MTA has been in a process of reform that did not end with the passing of the law on the Academy in 1994.

The MTA has been a central institution of the Hungarian S&T system for decades now. The organization, which, during the era of the People's Republic, was responsible for practically all basic research and a substantial part of applied research as well as development, was mentioned repeatedly by the Organisation for Economic Cooperation and Development (OECD) examiners in their report on the Hungarian S&T system (OECD, 1993, p.112). The main proposition of the examiners was that the Academy had too many functions. The conflict between advisory/lobbying functions and funding/management functions were singled out here in particular. The report does not, however, suggest the Academy should restrict itself necessarily to the typical functions of Western Academies of Sciences. On the contrary, it mentions the 'remarkable competence accumulated in a number of research groups' in the MTA's network 'for which it is responsible' and which it should preserve, notwithstanding a necessary restructuring of the organization (OECD, 1993, p.113).

These recommendations are comparable to the suggestions made by the International Council of Scientific Unions (ICSU) that also carried out an evaluation program of a series of MTA institutes in 1992. The ICSU examiners particularly mentioned the needs for a reform of the MTA's centralized structure, for renewed cooperations with universities, for a reduction of permanent positions and for 'a healthy balance between contracts with industry and fundamental research...in the medium- and long-term' (MTA 1993, pp.3).

Indeed, since the beginning of Hungary's transition to capitalism an internal reform process has been changing the structure and workings of the organization. Hungarian researchers living abroad are invited to take part in decision-making processes. The Academy is now less centralized than it was during realsocialism, allowing more freedom for research institutes. The voting procedures have been changed with the goal of opening the Academy up, making the internal election processes better visible. The supervision of the National Scientific Research Fund/Program (OTKA), has been handed over to the Ministry for Culture and Education – if only to be regained later. The advisory function to government has been regulated by the Law on the Academy (MTA 1994), under which the organization produces an annual report on the doings of the Academy and the state of Hungarian science and lays it before Parliament. More importantly, the Academy's independence is now guaranteed by law. The organization governs a number of R&D institutions, most of them in the realm of basic science. A few former institutes are now independent firms, engaging in a variety of different specialized services (Balázs/Plonsky 1994).

Yet, despite these promising developments, the organization has not been renewed sufficiently to face Hungary's demands at the end of the century. Indeed, this is not only an opinion taken at times by outsiders, but merely a reflection of the MTA's own reform efforts. Besides the two external evaluations by the OECD and the ICSU already mentioned, the MTA undertook a major self-evaluation program in 1992. All of the organization's research institutes were scrutinized. The data were analyzed by a MTA committee, which was to make recommendations for improvements in financing, i.e. expenditure cuts. The direct results of the process were negligible, however, since 'there were practically no evaluation guidelines' (Hangos 1997, p. 75).

At the same time the MTA had to face severe budget cuts, especially between 1992 and 1995. As a reaction, the MTA had to set free 35% of its staff personnel by 1994. At the end of 1996 an internal evaluation of the organization was on its way (again), which was (again) expected to result in information on the decision as to which units were to be downsized or cut and which were to stay.[20] As a result of the evaluation, a reorganization of the MTA took place in 1997, in the course of which 1200 members of staff were set free and a number of institutes merged. Of the MTA's 56 previous research institutes and laboratories, 14 were merged or integrated into universities. To cover the extra costs of this reorganization the government made million 600 HUF available for the MTA (OSI 1997/4, p.43). Yet, a year later, another evaluation was started by MTA president Ferenc Glatz, which was intended to downsize the organization once again. And again a committee consisting of several leading persons of the organization was to make decisions on a reorganization, and there is still no immediate end of the Academy's transformation in sight.

The Chemical Research Center of the MTA may serve here as a case study of the changes the Academy's research institutes have gone through. The Institute was founded as the Central Research Institute for Chemistry in 1954. The main scientific focus of the organization lay on researching the relationship between the structure and reactivity of molecules. In its heydays the organization consisted of 500 members of staff and was closely cooperating with the successful pharmaceutical industry of Hungary. Around 60–70% of the institute's revenues came from long-term research contracts with this sector of industry.

After 1989, the number of contracts with industry diminished rapidly, due to the restructuring processes these companies were in. In the early 1990s the pharmaceutical industry was finally bought up by multinational firms, which showed no interest in the creation of new drugs in Hungary. Service-like activities such as test series or the synthesis of a specific compound, which are more development- than research-oriented, were the only contracts, which could be gained by the pharmaceutical industry in the 1990s. The direct contact to industry, which before 1989 was maintained on a professional as well as on a personal level – personnel from the pharmaceutical industry earned postgraduate degrees at the institute – was severely disrupted. Efforts to make up for the losses by establishing links to international organizations and foreign firms proved difficult.

In 1998, after several rounds of downsizing, the institute consisted of roughly 120 permanent staff and 40 Ph.D. students. In the 1997/8 reorganization of the MTA research institutes, the Central Research Institute for Chemistry was merged with two other organizations, the Isotope and Surface Science and the Materials and Environmental Chemistry Institute to form the Central Research Institute for Chemistry of the MTA.[21]

As a result of all of this, the MTA's future role in the S&T system a decade after the beginning of the latest transition still is not clear. The government under Prime Minister Horn has decided to entrust the Ministry for Culture and Education with the supervision of the Academy. Therefore, for the first time since the founding of the People's Republic, the MTA is organizationally linked with – and theoretically even subordinated to[22] – another bureaucratic unit. These structural changes, in addition to the alteration of the strategic priorities for S&T as well as the emphasis which has been laid on applied science since the beginning of the transition period and the resulting creation of the applied science oriented Zoltán Bay Institutes has undoubtedly had an effect the MTA.[23] Nevertheless, the MTA is an influential organization in the Hungarian S&T system, which it has repeatedly proven. This has become obvious in the aftermath of the 1995 budget austerity measures, during which the Academy has been capable of halting the deterioration of its financial situation with a series of extraordinary budgets, including a total of HUF 900 million extra to be invested into strategic research over over a time period of three years from 1997–1999 and, as already mentioned, HUF 600 million for the reorganization of the research institutes in 1997.

Table 6.9 tries to condense the functions which the MTA has been featuring. On the x-axis the range of functions of the MTA are shown in an order trying to mirror an increase in power from the left to the right. On the y-axis the time span of the existence of the MTA is shown. The time is partitioned into periods during which the organization had fairly comparable functions to perform.

The dimension 'body funding scientific activities' needs a clarification. The Academy was cosponsored by the government for most of its existence. However, as has been mentioned, the significance of government support increased in time. In this respect the most significant events happened during the first half of the 20th century. The MTA lost most of its assets first to inflation during the early years of the 1920s, then, some twenty years later, to the Communist nationalization program. Despite the fact that property was granted to the Academy by the state after the fall of realsocialism, the organization has not been able to support itself on its own. Therefore the dimension 'body funding scientific activities' stands for the distribution functions of the MTA in its role as an intermediary institution in the S&T system.

Table 6.9 Crosstabulation of the Functions of the MTA, 1825–1997

	1825–1918	1919–1948	1949–1989	1990–1997
learned society with an honorific character	yes	yes	yes	yes
body funding scientific activities	yes	yes	yes	yes
influential interest group of scientific community	?	?	yes	yes
umbrella organization for basic research institutes	no	no	yes	yes
organization for the accreditation of scientific degrees	no	no	yes	no
advisory body to government	yes	no?	yes	no?
ministry for research and development	no	no	yes	no

The Academy of Sciences (MTA), together with the National Committee for Technological Development (OMFB), the Ministry for Education and Culture and the National Scientific Research Fund (OTKA), cosponsors the National Information Infrastructure Program. The goal of this effort was to reach international standards in fundamental communication infrastructure, which makes the various Internet services possible, by the end of 1997. In the first part of the program, which extended from 1987 to 1994, all research institutes and many research groups of the MTA, as well as the larger universities and higher education institutions had been linked to the Internet (MTA 12/1995, pp.6). Furthermore, by the end of 1997 over 300 high schools were linked to with the Internet. An extension of the original plan wants all Hungarian high schools to have access to the Internet by 2002 (OSI 1997/4, pp.43).

Since 1994 the Ministry of Education and Culture has been among the major actors of the S&T system (Abott 1994, p.6A). Most notably, it has been endowed with the supervision of the newly founded Higher Education Research Council, by making the Minister of Culture and Education the chairman. The task of this body is to reform the higher education institutions of the country. In addition, the ministry has been supervising OTKA for two years, named until 1995 the National Scientific Research Fund and now the National Scientific Research Program, and FEFA, the Higher Education Fund. FEFA has been established by means of loans of the World Bank for the purpose of modernizing Hungary's higher education institutions.

OTKA has expanded its number of programs and the size of its support. Although it was originally planned to be an additional funding possibility for basic research projects, it has now become an indispensable income possibility for researchers in a variety of disciplines. Due to the shrinking governmental support – the budget which the fund received in 1997 was about a third smaller than the one obtained in 1991. However, the demand of the underfunded Hungarian innovation system is just as large as before and therefore the actual amount of the individual grants is small. More precisely, the average amount of money actually granted is on average a third of the sum that was applied for. Since 1992 a program for the support of junior researchers has been in effect. This program has been established to prevent young scientists from becoming part of the brain drain. Also, a postdoctoral training program has been created (Pécsi 1993, pp. 93; OSI 1997/3, p.52).

Since 1996 – paralleling the reform of KMÜFA – OTKA has no longer upheld the status of an independent fund anymore, but has been transformed into a program, which is part of the 'Economic Development 'Fund'. Moreover, the Ministry of Education and Culture has had to give up its supervision of OTKA, which has been transferred back to the MTA.

The way both OTKA, the National Scientific Research Fund/Program, and KMÜFA, the National Fund/Program for Technical Development, have been managed has become subject to severe criticism. The small amount of money of the individual grants handed out by OTKA makes it impossible for the researchers to sustain their projects. In addition, the fund has been charged with having ineffective reviewing techniques with regard to the incoming applications.[24] Moreover, since the project proposals are written in the Hungarian language, only a small number of international reviewers have been included in the evaluation process; and, since the number of scientists in a small country such as Hungary is small, too, there is 'no real anonymity on the part of the reviewers, and the objectivity of the reviews is sometimes questionable' (Hangos 1997, p.77).

OMFB support has been criticized for not sharing risks adequately. As has earlier been said, large parts of KMÜFA funds are used for loans with a comparatively low interest rate. The problem in this context is, critics insist, that OMFB does not effectively share the risks of new investment by granting loans in most cases. Moreover, industrial partners often decide not to take their chances and therefore do not invest adequately in innovative projects.[25]

Another smaller fund was set up in 1992: The National High Priority Social Science Research Fund (OKTK), which used to be under the supervision of the Science Policy Council and which is now dispersed by the MTA. The objective of the OKTK is to fund projects which come up with research results directly usable for policy makers.

In fall 1992 the Zoltán Bay Applied Science Research Foundation was set up by government. The impetus for forming an applied research network came from OMFB. In 1993 two institutes had been initiated, a third followed in 1994. The institutional reference for this organization was the German Fraunhofer Gesellschaft, which in the German S&T system has the role of forming a linkage between academia and industry. The institutes went through a period of strong growth, with just over 60 employees by the end of 1994, but more than 130 employees in 1997, of which two thirds were researchers. In addition to this, around 20 Ph.D. students were working for the organization at that time. Of all the doctoral students which left the institutes by the end of 1997, around a third have found their way to industry.[26]

In a self-characterization the Bay Zoltán Foundation elaborated on its relationship to universities by stating that it wanted to
- involve university teachers in the institute's research and institute researchers in the university's teaching programs,
- utilize both university infrastructure and institute infrastructure for R&D carried out by personnel of both institutions,
- use the results of the university's basic research for further R&D, and
- give students the chance to engage in R&D (Bay Zoltán Foundation 1995, p.5).

In a later self-characterization the Foundation put more emphasis on the participation in international projects and the development of marketable products and technologies of international standard (Bay Zoltán Foundation 1998, p.3).

The Bay Zoltán institutes have already had their first successes. A number of innovations have been made, some of which have found their way to successful implementation: in Szeged in the area of environmental and bio-technology, in Miskolc in the fields of logistics and production technologies and in Budapest in material science. Funding for the institutes so far has come from industry (a third), from projects carried out for governmental organizations (a third) and from basic funding by the state (a third).[27]

Amongst all the professional organizations founded in Hungary during the first decade of the latest transition, one seems to be most outstanding, namely the 'Innovation Chamber', which was founded in 1990. In 1995, due to changes in legislation concerning a variety of interest groups, the organization had to alter its name to the 'Hungarian Association for Innovation' (MISZ). It is a non-governmental initiative – which in itself makes it a wondrous creature in the insecurity of a transition environment. The organization represents the interests of its members, which are companies, innovation parks and R&D institutions from the realms of universities, the Academy of Sciences and higher education institutions. At the end of 1995 the Association counted 240 members.

Moreover, via the National Business and Innovation Center the organization tries to offer services to companies. Via the Grand Innovation Prize and the National Scientific and Innovation Contest for the Youth the association makes an effort to instigate innovations, publicize them and create an innovation-friendly culture. The highly active organization is involved in a number of projects concerned with technology transfer and is now cooperating also with government organizations. The positive role of the Innovation Association has been widely acknowledged in the whole Hungarian S&T system.[28]

Another dynamic institution providing financial help to the Innovation Association is the Foundation for the Technical Progress of Industry (FTPI), which was established in 1990 by the Ministry for Trade and Industry and an industrial development bank, the 'Corvinbank'. FPTI's mission is to support the restructuring of industry and provide a steady influx of new ideas to the Hungarian production sector. To meet this end the organization has established a number of initiatives under its auspices, namely the Logistic Promotion Center, the Hungarian Quality Development Center for Industry and Trade, the Hungarian Technology Transfer Center and the Hungarian Association for Environmentally Aware Management. On top of this bundle of activities in key areas, the FPTI also offers information dealing with a number of innovation-related topics.[29]

Both, the FPTI and the Innovation Association are, amongst others, aiming at start-up companies, which have to overcome a number of hurdles in Hungary, of which the dearth for capital may be the most important. Although, in this respect, Hungary is better off than the other countries currently in transition from realsocialism to capitalism, a problem for start-up companies which has not been overcome, even at the end of the 1990s, is the lack of venture capital funds. It is questionable whether high risk funds could be provided by private banks under the constraint of the current capital scarcity. Consequently, the Ministry for Industry and Trade (IKM) together with private entrepreneurs, has set up two projects, IRCIL and Multinova Co. Ltd., both of which channel funds into start-up companies after investigating their chances for success (OECD 1993, p.131). These two and Covent, another government-backed fund, finance high-tech and R&D companies. In addition to this, OMFB supports small and middle enterprise (SME) start-ups. These initiatives, together with a number of foreign private and government sources, have resulted in a small venture capital industry, which by 1995 had reached a total estimated size of USD 300 million (Meth-Cohn 1995, p.8).

Indirect financing of R&D, as it is practiced in all OECD countries to varying degrees, is not very common in Hungary.[30] Inzelt notes that 'it is mainly foreign investors who enjoy tax breaks'. (Inzelt 1995a, p.32) Moreover, the financing of R&D is frequently subject to negotiations between investors and government. Direct instruments as loans and subsidies are regularly employed. The existing tax allowances – since 1997 20% of a firm's R&D costs may be deducted from taxable income – have been criticized by interview partners as being insufficient and especially hurtful to SMEs.

In order to raise funds and gain leading edge technologies, cooperation with other countries has been fostered by the Hungarian government. Linkages have been established in private, public, profit, non-profit, scientific, technical and political forms. Besides foreign direct investment, which is to be addressed later on, international programs are noteworthy in this respect: Hungary has succeeded in entering a wide variety of partnerships and programs of bilateral as well as multilateral nature. As far as bilateral agreements go, Hungary, together with Poland, turned out to be the main S&T Central and Eastern European (CEE) cooperation partner for a number of countries soon after the start of the latest transition. This is especially the case for the most important S&T cooperation financier, the FRG, which invested large sums into CEE S&T during the first half of the 1990s (Biegelbauer et al 1998). As far as multilateral agreements go, the country is involved in programs of international organizations like the World Bank, the EBRD, NATO and the European Union, to name only a few of the most significant. These agreements have an important role in the S&T system, since they introduce best practice procedures to the country in a variety of forms, extending from norms and standards, to forms of work organization and project management.

Hungary, during the era of realsocialism, was the CEE country with the highest concentration of its workforce in large companies. A broad societal consensus in Hungary exists stating that for a small open economy a concentration of economic power in a few companies is not advisable. Nevertheless, the deconcentration of the productive capabilities is advancing only slowly. Still, the percentage of industrial organizations with more than 300 employees fell from 86% in 1989 to 59% in 1992 (Inzelt 1994, pp. 141). These smaller companies frequently miss the financial potential or the know-how to engage into R&D.[31]

Another society-wide consensus in Hungary is concerned with privatization which has been identified as necessary for the country's development towards a market economy. There are, however, a variety of concepts and a great many ideas about the speed of privatization measures.[32] The actual privatization process has begun more slowly than expected, whereby different factors have been responsible for this slow advance of privatization; in this context, the shortage of capital in the national economy, the decision of the government to hold on to at least 25% of the stakes in many companies, including the industrial R&D institutes, and quarrels between different parts of the executive powers are reasons that have been cited (Chesler 1994, p.1/2).

Moreover, in October 1996 another possible reason for the moderate speed of privatization came to the limelight: corruption. The lawyer Marta Tocsik had been paid HUF 804 million (more than USD 5 million) as consulting fees for negotiating shares for municipal real estate. Since the 'fees' were apparently paid for an insubstantial amount of work, the investigation soon crystallized around the term 'corruption'. Having been in office for only two months, Minister for Industry and Trade, who also was Minister for Privatization, Tamas Suchman, had to resign. Suchman, who after Laszlo Pal and Imre Dunai was already the third Minister for Industry and Trade under Prime Minister Horn, was replaced by Szabolcs Fazakas.[33]

Whatever the reasons for the slow start of the privatization may be, at the beginning of 1994 only twenty percent of the formerly state-owned enterprises were sold (Inzelt 1994, p. 145). However, since 1995 the privatization has been accelerating. At the end of 1995 privatization had reached nearly 70% of Hungary's national economy and in mid 1997 state ownership had been reduced to less than 50% in 1.489 of the 1.698 joint-stock companies targeted for privatization in 1990 by what was then called State Property Agency (EBRD 1997, p.173). By the end of 1998 only a handful of state-owned enterprises were waiting to be sold.

The part of the S&T system which has been suffering most extensively under the slow progress of the privatization program is the sector of the industrial R&D institutes. As has been mentioned, a large part of industrial R&D was performed by specialized R&D institutes, which were supervised by several branch ministries during realsocialism. Around 40 institutes were responsible for high level industrial R&D, whereby the Ministry for Industry and Trade was the major supervising institution for these organizations. By 1995, already about a fourth of these organizations had been closed, most of the others had been downsized, in some cases dramatically so. These activities of the organizations which had survived until then had been reduced to service, trading and production, with only 10–15% of their activities being R&D related (Mosoni-Fried 1995, pp.777). Of the 17 purely industrial R&D-related institutes 5 were liquidated and only one was privatized during the first half of the 1990s. In 1997 two of the 11 remaining institutes were merged and handed over to the privatization agency AV Rt., while one further institute was liquidated. The other three institutes were to remain in state-ownership (Bouché 1998, pp.183).

One comparatively successful example of the otherwise ailing industrial S&T sector is the Autokut institute. The institute was founded in 1950 with the task of carrying out R&D for vehicle development. In its heydays the institute, which was a so-called R&D company, at the beginning of the NEM employed a staff of about 500 and developed engines, axles, gearboxes and other high value-adding machine parts. In 1988 this number had already shrunk to 100, a sign of the transition to capitalism, which had already begun before the fall of the iron curtain in Hungary. In 1994 the institute was transformed into a joint stock corporation, with the state as a shareholder of 25% plus one share. By this time the institute's number of contracts had decreased sharply, since the main customer, the Hungarian bus producer Ikarusz, had lost its largest market, the Soviet Union.

In 1998 the institute had sold a lot of its real estate and reduced its staff to 60. Its main income was generated through the testing and measuring of vehicle parts, the revenue and further development of an antijacknife device for buses and the development of engines for a Hungarian and a Polish company. A few of the dismissed engineers were still working in industrial R&D, but most of the former staff had retired. The future of the institute was still not secured, but talks were held with the privatized Hungarian truck producer Raba, which was interested in buying into the institute.[34]

In general, the statement holds true that of all parts of the national economy in transition, industry and the connected R&D system went through the most disruptive adjustment processes. Translated into data, the industrial R&D system, released between two thirds and than three quarters of its R&D personnel – depending on the source used.

After more than 10 years of transition from realsocialism to capitalism it is still difficult to generate reliable data on the development of the industrial R&D system during the transition period. The reliability of statistical data available on industrial R&D is lower than that on the overall national S&T system. After all, the changes for the industrial R&D system have been more pronounced than for the rest of the system. Judit Mosoni states that the systematic and continuous efforts to get an overview of the development of the industrial R&D system by the National Committee for Technological Development, OMFB, were canceled in 1991. She describes the Central Statistical Office's efforts a year later to question 641 economic units which, when they were established, had stated to have R&D as their main task as follows:

61 questionnaires were mailed back as 'not known', 439 units answered that they did neither research nor development in 1992. Appreciable question-naires (accounting reports) were sent back by altogether 22.6% of the

requested firms. They employed 1104 persons (non-permanent staff) for R&D work. (Mosoni 1994, p.6)

The data from this survey suggests that the actual R&D performed in industry may have been even smaller than the large number of personnel that had been laid off would indicate.

Another survey done by the firm Szonda-Ipsos also indicated a low level of R&D activity in industry. In this project initiated by the Ministry for Trade and Industry, managers from 2.000 industrial firms were polled. 57% of the firms had no R&D effort. 30% of the firm representatives said they had used in-house capabilities for their R&D, mostly consisting of occasional jobs for employed engineers. Only 20% of the firms had in-house R&D units. A mere 15% contracted R&D out to other institutions (Balázs 1994, p.8).

In 1994 another survey was carried out, this time by the Budapest-based Innovation Research Center (IKU). Out of 478 enterprises only 110 responded, after having been contacted by mail and telephone.[35] Perhaps the most interesting information provided, is the ranking of factors hampering innovative activity which was given by the companies contacted. This ranking is reproduced in the following table.

Table 6.10   Ranking of Factors Hampering Innovative Activity in Hungarian Companies, Number of Firms with Answers of 'Very Significant' and 'Crucial' (Total = 110)

Economic Factors:

Lack of Financial Resources	80
Innovation Costs too High	48
Pay-off Period too Long	40

Enterprise Factors:

Innovation Potential too Small	36
Lack of Information on Markets	18
Innovation Costs Hard to Control	18
Organizational Structure of the Enterprise	18

Other Reasons:

Legislation, Norms, Regulations, Standards, Taxation	28
Lack of Technological Opportunities	21

Source: Inzelt 1995, p. 14.

The primary factors identified as hampering innovativeness all appear to be linked to financing. The dearth for capital resulted not only from an undercapitalization of the companies, but also from a high inflation rate of 28% for 1995 and, subsequently, from high interest rates set by the banking sector. In addition, available funding from government decreased, just as the foreign direct investment (FDI) influx became smaller during 1994 (Business Central Europe 1995).

In 1997 the Ministry for Trade, Tourism and Industry (IKM) analyzed the industrial R&D sphere. Of 3.546 companies responding to the ministry's survey, 373 (10.5%) stated that they carried out R&D. Of these 213 companies, which could clearly describe their R&D activities, only slightly more than 50% had activities exceeding 1% of the companies' total turnover. In the final report of the survey, industrial R&D was depicted as slowly picking up towards the end of the 1990s, even if from a very low level.

In the course of the survey a closer look was taken at the effects of FDI on industrial R&D. It turned out that 11% of the firms maintained R&D contacts, 13% marketing contacts and 9% production/manufacturing contacts with multinational companies (IKM 1997). Despite the relative gains compared with the situation during the first half of the 1990s, these numbers still seemed rather low in the light of the otherwise extensive engagement of multinational companies in Hungary.

In fact, Hungary absorbed more FDI than any other formerly realsocialist country in Central and Eastern Europe. With the onset of the latest transition, a steady stream of FDI has begun to pour into the country. In October 1994, the German Audi company opened a Deutsche Mark (DM) 300 million motor-assembling factory in Győr, half an hour's drive from the Austrian border. Investments of a further USD 274 million are planned by the year 2000.[36] In April of 1996 Sony decided to invest USD 20 million in a 900 people greenfield plant in Gödöllö for the production of mini-stereo systems, color TVs, VCRs and CD players.[37] General Electric (GE), which bought Hungary's Tungsram in 1990, shifted European lightbulb production from other countries to Hungary, exporting as much as 90% of the Hungarian plant's total production in 1994 (Smart 1994, p.49). GE's investment was the second highest of a foreign company in Hungary with a total of USD 550 million (Wörgötter et al 1995, p.103), some of which was put into the R&D facilities of Tungsram. Nevertheless, it should be mentioned that the decision to keep R&D facilities in Hungary was made as late as 1995, five years after the initial engagement of GE.[38]

Altogether, by the end of 1996 USD 15,2 billion in foreign capital had been invested in Hungary, which was more than in the other countries currently in transition from realsocialism to capitalism. Half of this sum was invested in industry: 15% went into the telecommunications sector, 13% into the energy sector, 6% into the financial sector and 6% into the trade sector. More significantly, '(i)n 1995 98% of the country's 200 largest companies was majority foreign owned'.[39]

In general, the foreign capital inflow into Hungary exploded with the beginning of the transition. As can be inferred from table 6.11, the share of FDI as part of the total capital investment increased steadily during this time.[40]

Table 6.11   Foreign Capital in Hungary, 1989–1994

Year	Foreign Capital (HUF bn)	Foreign Direct Investment (HUF bn)	Share of FDI in Total Capital (%)
1989	124,4	30,0	24
1990	274,2	93,2	34
1991	475,6	215,0	45
1992	713,1	401,8	56
1993	1113,2	662,9	60
1994	1398,2	833,5	60

Source: Central Statistical Office 1996, p.19.

In 1994, German capital was responsible for 22% of FDI in Hungary, Austrian capital for 19% and capital from the United States for 14%. In a 1996 report, the Hungarian Central Statistical Office points out that

(I)n 1994, 39% of the total value added, 38% of the total sales receipts, and 38% of the total gross fixed capital formation was generated by FDI enterprises. This participation in employment is about 24% of the total number of employees in the corporation sector. It implies that their per capita turnover (61%), and per capita value added (67%) are higher than those of the total national economy. The average salaries are higher only by 25% compared to the enterprise sector. (Central Statistical Office 1996, p.20)

Partly as a result of the bad employment opportunities for Hungarian researchers at foreign-owned facilities, a discussion has been incited on the usefulness of FDI for the Hungarian R&D potential. For the Hungarian company structure FDI seems to have had favorable effects. 'Deep restructuring' of companies was led by Westerners bringing their business practices, which resulted in a stronger competition amongst Hungarian based firms this leading to leaner companies (Papp 1996). Moreover, capital investments, export structure and size of sales as well as other variables were positively influenced by FDI. In addition, privatization in Hungary frequently has been FDI led, as capital is missing in the country. This all might have an indirect effect on R&D, however, as has been indicated before, direct investment into R&D seems to be low. And just as the profits of foreign companies at times seem to find imaginative ways out of the national economy,[41] know-how might not stay in the country either. This phenomenon could be reinforced by a barely developed backward integration of the foreign-led firms.[42]

FDI can become dangerous for the development of a country when the foreign companies have enough leverage to buy monopolistic markets from governments in dearth for capital. David Stark points at cigarette manufacturers and, more important for industrial R&D, to automobile producers, who in the countries currently in transition from realsocialism to capitalism seek 'state subsidies, preferential credits, and strict import restrictions to reduce competition' (Stark 1995, p.15). There are indicators that this has happened in Hungary, too. Jan Stankovsky cites a number of credible sources for Hungary and other countries (Stankovsky 1996, p.134).

A study by Annamária Inzelt gives a mixed picture of the activities of the foreign firms in Hungary. The new foreign owners of large economic units in Hungary tend to speed up innovation processes, but do not invest heavily into Hungarian R&D. Furthermore, the multinational companies (MNCs) in general tend to spend less on R&D than the nationalized firms during the realsocialist era, but more than the Hungarian firms in the transition period.[43] This does not come as a surprise, given that MNCs – especially during the first years of engagement – primarily set up primarily assembly lines in Hungary. The parts used in the process are mostly imported,[44] although the situation slowly seems to change to the better.[45]

During the last years of the century, a number of firms have begun to invest into R&D facilities in Hungary, amongst them Elektrolux, Ericsson, Nokia and Siemens. It remains to be seen if these investments are of long-term nature.

A drastic example for the effects of FDI is the fate of the Institute for Telecommunication Research (TÁKI), which in the mid 1980s had a staff of some 1300, in 1996 consisted of 200 people and is scheduled for closure with the end of the 1990s. In the meantime a US-German consortium has invested USD 875 million into telecommunications – the largest piece of FDI in Hungary (Wörgötter et al 1995, p. 105). While this investment has brought a spur of activities in the telecommunication sector, finally leading to the much needed modernization of the Hungarian telecommunication infrastructure, it has also destroyed lots of indigenous research capabilities. However, the money previously spent by government on R&D in the sector seems to have a lasting effect. The personnel of TÁKI – including the people now working for the foreign-led companies – still hold contact to each other. The researchers know each other not only from TÁKI, but in many instances from their student time at the technical university Budapest in the 1960s and 1970s. Interactions between researchers extend from the exchange of information on funding possibilities to the lending of measuring instruments.[46] Thus, a network of Hungarian researchers has been formed, which develops and by itself stores knowledge. It will be interesting to see if this knowledge is going to have tangible effects on the S&T system, for example in the form of spin-off companies.

The Hungarian government is neither willing nor able to take over the financing of R&D from industry. In fully industrialized Western countries, private non-profit organizations (NPOs) at times have an important role in stepping in when governments fail to provide funding for R&D – especially in high-risk sectors.[47] Since 1987, following new regulations for NPOs, the number of foundations has been sharply increasing. Yet, although there is a vast number of NPOs in existence in Hungary – already by 1991 there were around 6,000 R&D foundations – only very few of them have an impact on the industrial R&D system. The reason for this is that most of them direct minuscule funds.[48] Since 1994 a law has been in effect that gives NPOs in general, not just those foundations essential for the innovation system, a new legal basis. It is, however, questionable if the impact of these organizations will be felt strongly in the industrial R&D system in the near future.

In comparison to the other sectors of the S&T system, the university system underwent comparatively slow and modest changes. Moreover, the funding for higher education was raised until 1995. Universities increased their efforts to engage into basic research again. The curricula of the universities were adapted to the changing demands of society. Noteworthy were especially the attempts of the universities to modernize and internationalize themselves via exchange programs, grants and lectures held in foreign languages, to name only a few initiatives.

During the last decades of the People's Republic documents addressing higher education were pushing for the reunification of the higher education sector. The goal was the creation of larger universities amassing a critical amount of teaching and research potential in Hungary. At the end of the 1980s the World Bank signaled that it would provide funds for the higher education system which Hungary had been asking for – under one condition: the reintegration of the universities and colleges. With the beginning of the transition the time was ready for making first steps towards this organizational restructuring. Following a number of position papers, but preceding the Law on Higher Education, the Antall/Boross administrations made a partial reorganization of the sector possible, without going all the way however: the higher education institutions were allowed to regroup and reorganize themselves into structures allowing for a larger degree of cooperation in areas as research, information infrastructure, curricula etc., but were not allowed to found common institutions. The Horn government rationalized the administration of the higher education subsystem by unifying the bureaucratic supervision in the Ministry for Education and Culture in 1995, but did not advance the organizational integration of universities and colleges further (Bessenyei/Melchior 1996, pp.49).

Another problem area which has been under discussion for several years is the question of entrance regulations for the university system. Each year entrance tests, complemented by lotteries, are conducted. There have been discussions about this legacy from the realsocialist era since the beginning of the latest transition period. At the center of the dispute is the question if it would not be more just to free the university entry from tests and lotteries. A similar discussion is focused on the question of tuition fees for university attendance. This hotly debated issue led to hefty student protests, the resignation of the Minster of Education and Culture Gabór Fodor and, finally, the introduction of fees in 1995. In the same year the tuition fees were part of the already mentioned drastic budget reduction plans introduced by the most-hated man of the Hungarian S&T community at that time, Minister of Finance Lajos Bokros. The austerity measures included a sharp reduction of university personnel – which, however, was followed by a 20% pay rise in order to meet the inflation rate.

The current number of students enrolled in university programs lies at around 130,000 – which is low in comparison to most European countries taking the number of Hungarians into account.[49] However, the numbers of freshmen are rising sharply. Whereas in 1989/90 20.000 students entered universities, in 1993/94 there were 35.000 and in 1994/95 already 41.000 – a doubling of new entrances in merely six years (Bessenyei/Melchior 1996, p.245).

Among the interest groups of the higher education sector we find the Hungarian Rector's Conference. Founded as early as in 1987, the organization has been a legal entity since 1991. Its goals are the representation of interests, the furthering of cooperation, expansion of international relations and development of curricula of the Hungarian universities. Another explicitly stated interest of the Hungarian Rector's Conference is 'to play a leading role in programs aiming at catching up to European higher education systems', as the self-description of the organization goes. To meet these ends the organization has established a secretariat together with the College Director's Conference and the Chair of Art University Rectors.

The Athenaeum Committee, which was set up with the General Assembly of the Academy of Sciences, is an example for the initiatives of the Hungarian Rector's Conference (Kosáry 1992). The committee was responsible for harmonizing the draft bills on the Academy and on higher education. In addition, it arbitrated the conflict between universities and MTA over the granting of scientific degrees. As an unwanted side-effect, a new conflict over the same issue arose between the state committee for the granting of the Candidate and Doctor of Science degrees and the Athenaeum Commission. The general function of the committee, however, was wider in scope. The original task of the Athenaeum Commission was to resolve a conflict with a long history: the dispute over resources,[50] that, since the creation of the People's Republic, frequently raged between the MTA and the universities.[51] Finally, the MTA agreed to support the universities with its resources. With the help of the MTA, the universities started a reform of their curricula and were authorized to grant Ph.D. degrees.

A still largely unaddressed problem is the provision of access to foreign literature for the personnel of the S&T system. While the interpersonal contacts of Hungarian scientists with the outside world are growing, it is getting harder for Hungarian researchers to gain access to foreign literature. Due to financial reasons it is becoming more difficult for libraries and information centers to provide information generated abroad. Several interview partners from different institutions central to the Hungarian S&T system pointed this problem out.[52]

Another problem barely tackled pertains to the financial situation of the Hungarian researchers. Since the beginning of the transition, academic researchers have suffered sizable losses of income, despite scattered attempts to ease their financial pains. For example, academic researchers received an extra wage increase of 20% in 1991, which however was vaporized by the 35% inflation rate. As a result, researchers have begun to leave even those institutes, which are not subject to closure.[53]

Depending on the research and industry sector in question, young researchers changing from the research sector to the industrial sector receive twice or three times the amount of money they can theoretically earn, if they are among the few lucky ones to get at least a stipend or research contract. And even if they get a grant or stipend, in all likelihood they will not be capable of staying in research, since there are practically no new jobs in S&T available. The brain drain of young Hungarian researchers and engineers is a problem that has been only imperfectly addressed by sponsorship programs administered mainly through the Fund for Higher Education (Felsöoktatásfejlesztési Alap, FEFA) and the National Research Fund (Országos Tudományos Kutatási Alap, OTKA).

6.4 Possible Explanations for the Differences

Since 1994, after four years of non-communist leadership of the country, Hungary's economy has been on a cautious upturn again. Much criticism has been brought forward insisting that Hungary's national economy had been faster on the rebound if other policies would have been employed. Is this true? What are the underlying reasons for possible policy failures? Why could some sectors of the S&T system develop seemingly better than others?

During the three phases of policies implemented since 1990 a variety of different problems can be seen. During the 'phase of euphoria' policy makers overestimated the possibilities of the market to recreate itself. In recent years social science has begun to understand the market as an institution requiring a number of socio-economic factors, in which it has to be embedded, to be a viable – and more effective – alternative to other forms of societal organization. As the experience of post WWII industrial policies in a variety of countries shows, these sets of factors often have to be created by governments.[54] Especially striking examples are Western Europe directly after WWII (Kregel et al 1992, pp.117) and the East Asian Newly Industrializing Countries in the 1960s and 1970s (Wade 1990).

Taking Hungary's experiences with governmental planning during realsocialism in consideration, it is understandable that the first government after realsocialism, headed by prime minister Antall, did not believe too strongly in direct interference with the market – and indeed would have had to face resistance from a variety of groups throughout society against more interference. The low rate of governmental funding for the industrial R&D system during the first two years of governance by the conservative coalition is explainable largely by this 'hands-off' approach. The effects of this policy on the S&T system, and even more so on the industrial R&D system, were severe as has been described in this chapter: a dramatic loss in personnel, a dearth for funding in all sectors and disciplines, a fall in investment activities.

It is essential to understand that these effects have been the consequence not only of governmental policies addressing the S&T system directly, but of a set of economic policies, originating from a specific ideology. This package of policies, while more gradualist than shock-like[55] in approach, still advanced a rather tight monetary course, aiming at the control of inflation and budgetary balance. High interest rates were one of the side effects of these policies, which led to a capital shortage and thus hampered the investments of corporations. The neoliberal regime in the import/export functions of the economic system, paired with the diminishing of subsidies, led to a reduction in export capabilities – at least in the short term.

Subsequently, all these political measures influenced the perceived needs in a negative way as well as hindered the actual capabilities of companies to invest into R&D. The necessity to secure the immediate survival of firms led managers to pursue short term strategies such as cutting down on R&D units and selling the productive assets – all of which severely hampered the innovation capabilities of industry (Balazs 1994, pp.6). In turn, the demise of industrial R&D had effects on the other parts of the S&T system, too. With all the imperfections in the cooperation between industry, universities and the Academy of Sciences (MTA) during the People's Republic, industrial R&D had become a critical factor for R&D activities both at the universities and at the MTA. Due to the shrinking budgets for industrial R&D, the research institutes of the MTA and of the universities have found life difficult, too. Worse, the linkages between universities and the MTA's institutes on the one side and industry on the other side have been severely damaged (Inzelt 1995, p.8).

During the following 'phases of frustration and realism' government began first to realize the necessity of crisis management and later the importance of strategic goals for S&T policy. However, the nature and amount of government support still were subject to criticism. Criticism on the way in which the National Scientific Research Program (OTKA) and the Basic Technological Development Program (KMÜFA) were managed has already been mentioned. Moreover, governmental policies have only gradually changed. This has partly been an effect of the general scarcity of resources during the transition. However, this argument is not really helpful in finding an explanation for the general fragmentation of policies in Hungary, including the hurtful non-synchronization of economic and S&T policies during the latest transition to capitalism. For an analysis of this phenomenon another framework addressing the different interest groups in the Hungarian S&T system is useful.

In any industrially developed country a number of interest groups can be identified in the respective S&T systems. One way to define the basic groupings is to distinguish groups through their respective field of work. Consequently, one could talk of three classes of interest groups, consisting of

– researchers primarily concerned with academic fundamental science in elite institutions with the function of centers of excellency,
– researchers primarily concerned with academic fundamental and applied science as well as teaching in higher education institutions,
– researchers primarily concerned with applied science and experimental development in intermediary organizations and industry.

These interest groups have their corresponding partners in government, so that for example interests of researchers in industry are most likely to be represented in government by some sort of ministry for industry, international trade or economic affairs, in the Hungarian case the Ministry for Industry, Trade and Tourism.

In most systems a number of these interest groups have additional representation in the form of professional organizations such as chambers or associations of some kind, in the Hungarian case the Association for Innovation, primarily representing a number of groups concerned with some form of industrial R&D, or the Rector's Conference, representing the higher education institutions. In a number of countries all scientists are represented in a single organization, which is regularly captured by researchers primarily concerned with fundamental research, often from a narrow subset of disciplines, as these have more interests in common than the scattered interests of researchers in industry. In the Hungarian case this might be said about the Academy of Sciences to a certain extent, which, although originally conceptualized as purely basic research oriented, also has had units working on industrial R&D for a long time. Since the early days of the People's Republic, the Academy has been dominated by the natural and technical sciences, which stands in stark contrast to the main emphasis on social sciences and humanities, found bevor 1949.

In the course of time these interest groups have built, have captured or have been captured by existing organizations in most countries. In the case of Hungary the institutional structures corresponding to the groups identified are:

- the Academy of Sciences and its research institutions for the 'elite group',
- the universities and other higher education units for the 'higher education group',
- a few scattered institutions, which often are either new, like the Bay Zoltán Institutes, or have been diminished in size and number such as R&D institutes, sometimes belonging to the branch ministries, for the intermediaries and the 'industrial R&D group'.

Corresponding governmental units are

- in the case of the Academy: a minister for science until 1994, since then personal linkages of leading personalities often stemming from political relations,[56]
- in the case of the higher education institutions: the Ministry for Culture and Education and to a minimal extent the National Committee for Technological Development OMFB,

— in the case of industrial R&D, especially for the Zoltán Bay Institutes: OMFB as well as the Ministry for Trade and Industry and in some cases branch ministries.

These relationships also can be characterized as patron-client groupings, in which the governmental units extend their patronage over the clients, the R&D institutions. Two forms of patron-client relations seem possible: one in which individuals and one in which groups are partners in the relationship (Tarkowski 1981, p.181). This seems to hold true for economies in transition, too.

The relations between institutions and key personnel at both government and R&D level are exchange-oriented. While the governmental institution offers patronage in the form of representation in the struggle for funding, the R&D institution offers to be part of the clientele legitimizing the very existence of the bureaucratic patron. Table 6.12 makes an effort to stylize the relations between patrons and clients in Hungary's S&T system.

Table 6.12 Patron-Client Relationships in the Hungarian S&T System, 1990s

Subsystem	Client	Patron
Elite Function	Academy of Sciences	Minister of Science (up to 1994), Strong Personal Relationships
Higher Education	Higher Education Institutions	Ministry for Culture and Education
Industrial R&D	Industrial R&D Institutes, Intermediary Institutions	National Committee for Technological Development, Ministry for Trade and Industry, Several Branch Ministries

If the relationships between governmental and R&D institutions are analyzed this way, it is interesting to notice that a straightforward relation between a single patron and a single client exists only in the case of the higher education institutions, which, not coincidentally, are the 'winners' in the gamble for resources during transition.[57] The Academy, which has been without an institutionalized patron since 1994,[58] but not without (a) personal one(s), can be identified as a 'loser', but in comparison to the third group, the industrial R&D institutes, has had to deal with modest losses only. The third group was the clear 'loser' in the game for resources. Not coincidentally, the industrial R&D institutes were the only group with a larger number of patrons, which were endowed with more or less leverage, depending on the sometimes rapidly changing situations they were faced with during transition. As a result of the patrons' conflicting interests, the industrial R&D institutes are pushed around during the transition period. This clear division in 'winners' and 'losers' can be risked, because in times of resource constraints, as they are to be found during the transition period, the struggle for funding becomes a zero-sum game: Some win, others lose, but in the face of diminishing resources there are more losers than winners.

Apart from the fact that the struggle for resources in this context was quite tough, at first glance the groupings of interests in the Hungarian R&D system are comparable to what one can find in other R&D spheres, too. However, if one looks a bit closer, it becomes clear that four decades of realsocialism in Hungary have left their carvings in the system. Especially the previous compartmentalization of R&D during the existence of the People's Republic has led to weak linkages between the different interest groups. Due to the fact that the former organization of society, despite the efforts of the NEM, hindered the interactions between the Academy of Sciences, the universities and industrial R&D institutes and company units for a prolonged time period, only quite imperfect communication structures were upheld. This has a detrimental and negative effect on the relations between different interest groups of the S&T system, who all have difficulties in communicating with each other. Efforts to overcome these rifts, such as the Athenaeum Project, which should foster cooperation between universities and the Academy of Sciences, have had modest effects so far. This again, might be explained by the dire financial situation that Hungarian S&T is in, creating a situation of reiterated zero-sum games and of the structurally conservative S&T institutions, which are missing the influx of new ideas embodied in young researchers.

This scattered structure of the industrial R&D grouping led to a minimal interest representation of this group during the transition. In addition, the Ministry for Trade and Industry, responsible for most industrial R&D institutes, was weakened during the conservative Antall and Boross governments. During this time, from 1990–1994, the National

Committee for Technological Development, OMFB, was strengthened. OMFB, however, did not directly represent the industrial R&D institutes. Even its funding activities were scarce, as it feared that the industrial R&D institutes would not be able to pay loans back (Mosoni-Fried 1995).

This weak interest representation might be one of the reasons why the industrial R&D institutes were to be privatized via the State Assets Management Company (ÁV Rt.), with the state keeping a strategic share of 25% plus one vote of the organizations. This decision of government to keep its fingers in each pie – quite understandable from the standpoint of industrial policy – did little to entice any prospective owners to invest large sums into the organizations, of which they would have had only partial control. This resulted in a lengthy privatization process confirming evidence from other privatization cases that a prolonged process of transformation for organizations is harmful to most of them for a variety of reasons: no investments are made in the meantime, decisions can be taken only in lieu of a short term horizon and assets are sold off to ensure the short term survival of the institutions. Efforts on the side of the branch ministries, especially the Ministry for Trade and Industry, to influence this process were not effective.

In the framework of the new Horn coalition government the Ministry for Trade and Industry was clearly strengthened, partially at the cost of OMFB, which lost leverage in the form of competencies and funds as well as its status as a ministry of its own. Nevertheless, as has been shown before, discussions about the future of the remaining industrial R&D institutes have outlived the Horn government, without a clear decision being in sight by the end of the decade.[59]

The second organization identified above as a 'loser' in the struggle for resources is the Academy of Sciences. As has been pointed out, the institution has been weakened to some extent by the changes in governance structure in 1994: in the Horn administrations there was no Minister for Science anymore. This was not a huge loss in terms of influence as the Minister for Science had only had a small office and no corresponding ministry and therefore had not been too powerful in the previous governments. More significantly, a widespread feeling of insecurity has been preserved among the personnel of the Academy which is due to the provision that Parliament decides annually on the sum to be allocated to the organization, whereby the actual budget levels often wildly goes up and down.

This feeling of insecurity on the side of the Academy, formerly a powerful interest group representing the centers of scientific excellency of the country, was strengthened when it became clear that in a capitalist Hungary the Academy would not be the only center of excellency anymore. On the contrary, the institution was lessened in importance in comparison to the one group of institutions employing the researchers concerned with the same range of activities as the Academy: the universities.

Now that a number of reasons why Hungarian S&T policies employed during the 1990s were less than optimal have been identified and now that the question as to why some sectors of the S&T system have fared better than others has been answered, it is time to turn to an evaluation of the 1990s' S&T policies, which would not have been possible prior to clarifying these two issues. To facilitate such an analysis, a regional point of view of this question shall be dealt with in the next chapter.

Notes

1   Hungary Report 2.8, 12–08–1996, 'New Stats Outline Export Picture', http://www.isys.hu/hrep/.
2   Newsletter, Hungarian Foreign Trade Statistics 1997, 1988/2 (10).
3   During the realsocialist system the highest degree universities could grant was what is generally referred to as the 'small Ph.D.'. The MTA, however, would grant for further studies first the 'Candidate of Science', similar to the Western Ph.D., then, after further publications, the 'Doctor of Science', comparable to a Western European 'habilitation'.
4   This was accomplished through the 'Law LXXX of 1993 On Higher Education', approved by Parliament on July 13, 1993.
5   Compare with the changes in the S&T system of the former GDR, where Werner Meske of the Wissenschaftszentrum Berlin reported a loss of 70% of the researchers of the system – already by 1993.
6   Patent statistics should be taken with a grain of salt, due to the peculiarities of such numbers, which are often contingent on national specifica. However, the sheer size of the decline of patenting activity is telling.
7   Compare the three phases considered here with the three time periods distinguished in: Wedel 1994, pp.14. Jozsef Imre similarly finds three stages of transition, 'fragmentation – stabilisation – reintegration'; see Imre 1998.
8   This misbelief was fostered by the policies of different international agencies, which assumed a spontaneous resurrection of the market from the ashes of the realsocialist planned economies. A detailed analysis of the problems arising in

the transition economies from laissez faire politics has been given by Kregel/Matzner/Grabher 1992, see especially pp.112.

9   Of course, the organizational structures were already existing. OMFB, from 1990–94 the 'technology ministry', was a governmental agency before, supervising the National Office of Measures, the National Patent Office and the National Bureau of Standards. The 'science ministry', that existed from 1990–94, consisted of a small office and the restructured Science Policy Committee, headed by a minister without portfolio.

10   For an analysis of the Cold War reflexes more or less well hidden in the reports of Austrian media on the CEECs' governing socialist parties, see Pribersky 1996, p.23.

11   See for example the 1997 regulations for taxation in Hungary, with the highest tax brackets getting dramatically less taxed than before, whereas the medium brackets have to pay only slightly lower taxes than previously, but had their social security payments raised. These regulations could have been introduced only by a socialist party with a truly pragmatist background. See, 'Ungarn's Premier gelobt: Keine weitere Belastung', in: Der Standard, 09–01–1997, p.15.

12   As can be inferred from the funding strategies of the Western European governments as well as international institutions, with respect to the prospects of reform the West was on the same emotional roler-coaster as the East of Europe (Biegelbauer/Giorgi/Pohoryles 1998). This becomes especially clear in the case of the reunification of the two Germanies, which has been led by unrealistic expectations on both Western and Eastern sides. People, who cautioned the high hopes of Germans, were disregarded as being indecisive 'Zauderer'.

13   Which, after all, is consistent with neoclassical economics, too. For example OMFB/IKM/Ministry of Finance 1993, especially pp. 58.

14   An interesting saying on the matter popular among university people in 1995 went 'Hungary has survived the Osmans and the Mongols. We shall survive this crisis too'. Another variant replaced the word 'crisis' with the word 'Bokros', who was the then finance minister. On the saying and the situation of Hungarian higher education in 1996, see Kálmán 1996, p.28.

15   Different figures have been provided about the actual numbers of university staff set free in 1995. The Ministry for Culture and Education might underestimate the actual figures as much as universities might overestimate them – by and large the figure of 15% seems to be credible.

16   However, the subject had not been taken seriously by students (and perhaps not by teachers, either) for quite a time. A number of former students have assured me that they basically spoke no Russian, although they officially had

to pass a number of tests. Some have advanced the view that this in fact was a kind of protest against the political system.

17 Schools in German speaking countries with a curriculum concentrating on a specific technical field, as communication technologies or electrical engineering, attended by students at ages 14–19. These schools aim at enabling students to be mid-level engineers with the title of 'Ingenieur'.

18 Whereas a senior expert at OMFB during an interview has indicated that KMÜFA loans have to be paid back only in the case of business success, a Hungarian social scientist doing research on the workings of the innovation system criticized the 'hard' character of KMÜFA's loans. Moreover, another public official stated that while the loans of OMFB may bear no interest, they have to be paid back by law. A certain flexibility, according to the interview partner, can be provided with regard to the extension of the period before the actual payback has to begin.

19 The R&D funds of the branch ministries – about which it is hard to get information – have been largely abandoned during the latest transition period in favor of KMÜFA.

20 In fact, two interview partners with an intimate knowledge of the organization have been very critical about the reform process of the MTA, which was identified by them as a non-reform. Despite all the working groups' and committees' and international organizations' recommendations the actual results of the process are negligible since the same groups of people as during the start of the reform effort still are governing the institution, according to the two interview partners.

21 Informations were obtained through personal interviews with institute staff and social scientists, as well as from the new institute's self-characterization 'Central Institute for Chemistry of the MTA', 1998.

22 This regulation may have reenforced the negative climate between the MTA and the Ministry for Education and Culture: the then president of the MTA, Domokos Kosáry, and the then Minister of Education and Culture, Gabór Fodor, officially never had met – despite the above regulation, which would have necessitated regular meetings between both.

23 Not surprisingly, most people in the Hungarian S&T system seem to expect the MTA's influence to decrease further.

24 See Balázs 1995, p.648. More criticism on OTKA for the time before the latest transition in: Bessenyei/Melchior 1996, p.231.

25 Balazs 1994, p.25; on risk-averting behavior amongst Hungarian entrepreneurs see also Inzelt 1995b.

26 This information provided by an interview partner in the Bay Zoltán institute seems to be comparable to results of similar programs in countries as diverse as Austria, Slovenia and the UK. What is different, however, is that industry

in Hungary pays much more in comparison to academia (around 200–250%) than is the case in the other three countries mentioned.

27   These figures are internationally comparable, too, since for example the German Fraunhofer Institutes or the Austrian Research Center Seibersdorf have a similar funding structure.

28   In fact, in all my interviews I did not meet one person, who seemed critical about the organization. In an environment such as the Hungarian S&T system, this is a rare coincidence.

29   Information from the Foundation for the Technological Progress of Industry's self characterization (brochure) and several interviews with director and staff of the Foundation.

30   It seems that a systematic usage of certain policy tools such as tax allowances and fiscal incentives of all sorts have been considered, but not given high priorities.

31   An indication for the SME's missing R&D capacities was the statement of an interview partner, who pointed out that the information infrastructure of the public organization for which he was working had been used extensively by the personnel of the large nationalized companies during realsocialism. At present the interest of companies was picking up 'very slowly'. The main reason for this, according to the interviewee, was that 'the newly founded SMEs have more basic problems to care about than collecting information'. However: without information no R&D is possible.

32   For an overview of a variety of concepts see Mihályi 1993, pp. 84.

33   See for example Hungary Reports 2.9, 16–08–1996 ('Industry and Trade Minister Dunai Resigns'); 2.19, 29–10–1996 ('Supervisary Board Blasts Suchman') and 2.20, 04–11–1996 ('Tocsik Skips Parliamentary Hearing', 'Csiha, Fazakas Appointed Ministers', 'Blame Pinned on Top Brass at APV. Rt'.), http://www.isys.hu/hrep/.

34   Information stems from personal interviews at the institute and with social scientists.

35   Inzelt 1995; given the efforts and experience of the social scientists at work this is a rather low turnout pointing at the unresponsiveness of firms.

36   Oltay 1994; Business Week gives somewhat smaller numbers, USD 195 million already invested and USD 280 million investment planned; see Smart 1994.

37   Hungary Report 1.45, 08–04–1996, 'Sony Invests $20 million in CD Factory', http://www.isys.hu/hrep/.

38   Earlier in 1995 an interview partner pointed out that the investment of GE had resulted in only two more Hungarian researchers. However, it is difficult to attain actual data on the multinational company's effort, since the director of

the R&D department in 1996 was not willing to provide information not even regarding the number of employees working for him.

39   Hungary Report 2.21, 11–11–1996, 'Ft 15 Billion invested in Hungary So Far', a resume of an article published in the Hungarian daily Nepszabadsag on 06–11–1996, p.10; see also Mosoni 1998, pp. 171.

40   For information on the 'Top 30 Corporate Foreign Investors', see the article in: Budapest Business Journal, November 25 – December 1, 1996, pp.8.

41   Papp 1996 cites a Kopint-Datorg study coming to this conclusion on p.42.

42   Besides the at times low local content level of foreign led firms, the fact that these firms import even more than they export points at a low level of backward integration: imported products and materials often are used for the assembly (transplants!) and production of goods, which are exported again.

43   Inzelt 1994, pp. 154; Judith Mosoni takes a similar stance in Mosoni 1998, pp. 179.

44   However, it is reasonable to assume that MNCs use the expenditures on R&D more effectively than did the nationalised industrial companies of the realsocialist era. Therefore the amount of money spent on R&D in Hungary by foreign MNCs is not a sufficient criterion for a judgement of the impact the companies have on the Hungarian R&D effort.

45   An interview partner estimated that the Suzuki plant in Hungary already had around 30 local subcontractors by the end of 1995. In addition, currently efforts are discernible to raise the local content level of goods produced in Hungary. Compare also with Czaban/Henderson 1998.

46   The information on the former TÁKI researchers originate from a Hungarian social scientist married to one of the former R&D personnel of TÁKI.

47   For an account of the role of the German NPOs see Campbell 1993, especially pp.8.

48   More than half of them had assets below one million Forint (which, in 1994 was approximately USD 13,000)! These data are from Darvas 1995.

49   A variety of figures can be found addressing the numbers of students and teachers in Hungarian universities. As there are differences between the Western European and the Hungarian system in regard to types, nature and length of studies, the total number of students depends on the scope of the term 'student'. Quite likely seem numbers of around 127,000 students and 17,700 faculty members (Darvas 1995). The ratio of university teachers to students is surprisingly low – between 1:4 and 1:8 (depending on the assumed total number of students), if all teaching personnel is included. For information on the Hungarian university system see for example Halasz 1989, Bachmaier 1991, as well as Bessenyei/Melchior 1996.

50   This was done partially. Despite the work carried out by the committee, individuals from both universities and the MTA have indicated in

conversations that their organizations may have lost clout in the S&T system's governance in an unduly fashion.

51   An interview partner working for the MTA pointed out that some of these conflicts seemed to have been politically generated, along the lines of Julius Caesar's dictum 'divide et impera'.

52   Both the directors of the OMK and OMIKK, the Central Technical Information Center and Library, as well as a senior staff member of the Central Library of the MTA stressed the severe financial problems which the institutions they worked for had to face. See also Herman/Stubnya 1994.

53   For a detailed account of the brain drain and in general 'Changes in the Personnel and Financing [of] R&D', see the equally named section by Tarnóczy 1995.

54   For quite differing accounts compare Zysman 1983, Hall 1986, Katzenstein 1985, Hollingsworth/Streeck, 1994. Even in the US, which for a long time during the post WWII era was a country with a strong disbelief in industrial policies, administrations during the 1980s became more active in the area of industrial policy. This is even more true for the Clinton/Gore administration. See, for example, Biegelbauer/Filzmaier 1996. In this respect, discussions about the national information infrastructure, precompetitive research consortia or the government partnership with the three big car manufacturers for building the '80 miles per gallon car', are exemplary for recent achievements; see for example Wald 1994, Field/Frank/Clark 1997.

55   Ever since the beginning of the latest transition of the CEECs an intense discussion has been waged between disciples of a slow 'gradualist' transformation of the economy from a centrally planned to a capitalist system and the people adhering to the rapid 'shock therapy'. The most prominent and controversial preacher of the shock-therapy has been Jeffrey Sachs from Harvard's Business School. An equally controversial person on the other side of the trench has been Roberto Unger, also from Harvard University.

56   Members of the Academy frequently were part of realsocialist governments. There has been an intimate relationship between parts of the Academy of Sciences and the political regime.

57   Given that one can talk of 'winners' in a situation of serious resource constraints. After all, the universities have lost personnel, too, even if less than the other two groups.

58   Specifically since 1994 the MTA has been faced with budget cuts.

59   Worse, the Horn administration decided to keep 51% of most industrial R&D institutes – even 25% more than the Antall/Boross governements had been insisting on. For a highly informative analysis of the industrial R&D institutions' situation see Mosoni 1995.

7 Presence and Future: Hungary and the other CEECs

7.1 Hungary's Economy in the Region

Due to its small open economy Hungary's future is dependent on the future of its neighbors, Austria as well as the Central and Eastern European Countries (CEECs), who are in the process of transition from a centrally planned to a capitalist system. Two trends are discernible in the relations amongst the CEECs in transition to capitalism: one of cooperation and one of competition. Attempts towards an economic cooperation have been made with the Central European Free Trade Association's (CEFTA) initiative[1] involving the economically most developed CEECs, the Czech Republic, Hungary, Poland, Slovakia and Slovenia. The CEFTA agreement has been slowly but steadily advancing in its main task to reduce constraints on mutual trade.

This set of countries with a vicinity to Western Europe, a common cultural heritage stemming from the Austro-Hungarian Empire and comparable economic and social problems is in a race against time. As they are economically dependent on the EU area, the Czech Republic, Hungary, Poland, Slovakia and Slovenia have to integrate themselves as soon as possible into the Western European context. Most of the CEFTA countries, with the notable exception of Poland, which features almost 40 million inhabitants, are small open economies, which can only survive by exporting goods. Policy makers in all five countries have recognized this need and address it by their repeatedly stated willingness to enter the EU as soon as possible – most likely during the first decades of the oncoming 21st century.

This, however, to a certain degree predetermines the relationships amongst all the CEFTA countries, but especially the CEFTA Five, which is what the most developed and competitive CEECs are called, namely the Czech Republic, Hungary, Poland, Slovakia and Slovenia, all scheduled for the first round of EU accession. And indeed, this very expectation that the sooner the CEFTA Five have established close (and exclusive) relationships to Western Europe, the better the chances for their economies to thrive, hinders the CEFTA agreements to be extended and even leads to reoccurring disputes between the CEFTA member countries. Insofar, the relationship between the CEFTA Five is at least as competitiveness-oriented as it is cooperation-oriented.

A second clue for the competitive nature of the five country's relations is provided by the center-periphery notion.[2] Being placed on the semiperiphery of one of the earth's economically highest developed regions, Western Europe, the five countries' trade is dominated by their Western neighbors. As a result the larger part of the CEFTA Five's trade is not directed towards each other, but towards Western Europe. Accordingly, the common interests of the five countries are influenced by the politics of the EU, which tends to preserve its own interests.

Nevertheless, a comparison of CEFTA Five data offers a possibility to evaluate the success of the Hungarian transition reforms in comparison to similar efforts of the other countries. Precisely this shall be done on the following pages: detrimental comparative economic data of the five national economies are going to be presented, analyzed and compared.

Table 7.1 presents the GDP per capita rates of the Czech Republic, Hungary, Poland, Slovakia and Slovenia for 1997. Slovenia, the by far richest region of former Yugoslavia, is leading, followed by the distant second Czech Republic, then come Hungary, Slovakia and Poland. Since 1990 the only change in this ranking has been the overtaking of Hungary by the Czech Republic in the aftermath of the austerity measures of the Horn government in 1995 – a situation, which might be reversed again in the face of the high growth rates of Hungary and the sluggish growth of the Czech Republic at the very end of the decade.

Table 7.1   GDP per Capita of the CEFTA Five in USD for 1997

	Czech Rep.	Hungary	Poland	Slovakia	Slovenia
GDP/Capita	5.050	4.462	3.512	3.624	9.101

Source: Business Central Europe, 'The Annual', 1998/9.

Table 7.2 displays the GDP growth rates of the CEFTA Five for the time span from 1991–1997. As can be inferred from the table, Hungary's economy finally had bottomed out in 1994. In 1994 and 1995 the country featured low growth rates, but for the rest of the decade the growth rate is expected to stabilize at a high level. The table is also an indicator for the dependency of the CEFTA Five on Western Europe. The main reason why the growth rate for 1996 slowed down for all five countries is that the economies in Western Europe showed a short-lived downturn.[3] The domestic markets of the small open CEECs are not strong enough to make up for the losses in their main foreign markets in the West.

Table 7.2   GDP Growth of the CEFTA Five, 1991–1996 (% Change)

Country	1991	1992	1993	1994	1995	1996	1997
Czech Rep.	−14.2	−7.1	−0.9	2.7	6.4	3.9	1.7
Slovakia	−14.5	−7.0	−4.7	4.8	6.8	6.9	6.5
Poland	−7.0	2.6	3.8	5.2	7.0	6.1	6.9
Slovenia	−8.1	−5.4	1.3	5.3	4.1	3.1	3.8
Hungary	−11.9	−3.0	−0.8	2.9	1.5	1.3	4.4

Source: Kopint Datorg 3/1998, p.182.

Note: Data for 1997 are preliminary figures.

The changes in industrial production, still the most important part of GDP for economically developed economies, are reflected in table 7.3. In 1997 the industries of the CEFTA Five were blossoming. However, the losses in productive capacities relative to the 1990 levels were still spectacular for some countries. For the first time since 1989 Hungary began to feature positive industrial production rates in 1992, a year after Poland – which, however, had already begun to featur deep losses in industrial production figures earlier. In 1994 Hungary was second only to Poland in the efforts to exceed 1989 production capacities. This situation is not expected to change for the remainder of the decade, with Hungary slowly gaining on Poland.

Table 7.3   Percentage Changes in Industrial Production of the CEFTA Five for 1997 and Year of Lowest Production Since 1990 (in Brackets)

	Czech Rep.	Hungary	Poland	Slovakia	Slovenia
Basis 1990	−15.7 (1993)	−6.3 (1992)	−38.8 (1991)	−29.5 (1993)	−17.9 (1993)
Basis 1993	21.9	28.7	46.8	20.6	10.5

Source: Kopint-Datorg 3/1998, p.182.

The next table details out the rates of industrial production for the period of 1991–1997. Hungary, paralleling the slower growth of GDP during the same time span, fell behind the other countries in 1995, only to show stronger growth since then. Nevertheless, Slovenia featured an even slower growth rate of industrial output in 1995 and the slowest growth rates since then. Slovenia, however, had started out from a much higher level of per capita GDP and industrial production. Again, the dependency on Western Europe is indicated by this development: Slovenia, the smallest country in the sample with barely 2 million inhabitants, is hurt most by the weak demand in its main export markets in the West. Moreover, Slovenia is not capable of compensating these tendencies through the home market – domestic demand rose throughout 1996 (this included household incomes, investment spending and public expenditure) (Kopint-Datorg 2/1996(5), p.70.), with basically no effect on industrial production and GDP.

Table 7.4   Growth of Industrial Production of the CEFTA Five, 1991–1997 (% Change)

Country	1991	1992	1993	1994	1995	1996	1997
Czech Rep.	−24.4	−7.9	−5.3	2.3	8.7	6.4	4.5
Slovakia	−25.4	−14.1	−10.6	6.4	9.0	2.5	2.7
Poland	−11.9	3.9	6.4	11.9	9.4	8.3	10.8
Slovenia	−12.4	−13.2	−2.8	6.4	2.0	1.1	1.0
Hungary	−16.6	−9.8	4.0	9.6	4.6	3.4	11.1

Source: Kopint Datorg 3/1998, p.182.

The following table shows the shares of value creation in the different industrial sectors for 1991. In order to create a framework for the CEFTA Five, data of the Western neighboring economies of Austria and Western Germany (Germany in its pre-reunification borders) have been added. Western Germany, the most developed economy in the sample, has an industrial structure markedly different from the other countries on display. For Western Germany the share of textiles and clothing on the total value creation in manufacturing is the lowest and the share of production of machinery, electrical machinery and vehicles the highest of all countries. Austria ranks between the CEECs and Germany, with the second-lowest share of value creation in textiles and clothing as well as machinery, electrical machinery and vehicles. Hungary is next, followed closely, however, by Poland in the two sectors, with Slovenia creating an astonishingly high share of 16% of its value in manufacturing with light industry, both foodstuffs and textiles and clothing. From the standpoint of the classification of production output, Hungary's economy is closest to the manufacturing structure of the Western countries in the sample among the CEECs included here.[4]

Table 7.5   Value Created by Sector in Shares for 1991 (in%)

Country	Chemical Products	Food-stuffs	Textiles, Clothing	Machinery, Electrical Machinery, Vehicles	Other Manufac-turing
W.-Germany	12	10	4	41	33
Austria	7	16	6	28	43
Hungary	14	10	8	26	40
Poland	7	21	9	26	37
Slovenia	11	15	16	21	37

Source: World Bank 1994, p.211.

Note: Information on the Czech Republic and Slovakia, back then still the CSFR, are missing, as they have not been included in the World Bank data set.

A comparison of the data presented in table 7.5 with the information contained in table 3.5 is also highly interesting. Leaving England aside and comparing Germany, Austria and Hungary, Germany had the by far highest percentage of its manufacturing sector in heavy industries, with Austria displaying a quite even distribution between foodproduction, textile, leather and clothing production, and heavy industries. Hungary, however, had a strong foodproduction, a rather small textile, leather and clothing sector and exactly the same share of heavy industries of the total manufacturing sector as had Austria – if on a clearly lower level. Almost 80 years later Germany still featured the by far largest heavy industry sector of the three countries, with both foodproduction and textiles, leather and clothing diminished. Austria had increased the relative importance of its heavy industries in comparison to its other manufacturing sectors. Hungary, equally, had enlarged its heavy industries, largely at the cost of foodproduction. Nevertheless, Hungary has restructured its manufacturing sector on a clearly lower level than the other two countries.

Another interesting indicator is the export specialization of the CEFTA Five with respect to the technology content of the products. In fact, in the mid 1990s, in comparison to the OECD average, the specialization patterns of the CEFTA Five were all biased towards labor and resource intensive industries, with human capital[5] and technology intensive production largely underrepresented. Despite these points they have in common, it is interesting to note that the specialization patterns are not static, but are subject to change. While Poland had specialized even more on labour and resource intensive industries, Hungary and the Czech Republic during the latest transition to capitalism have shifted towards human capital intensive industries.

Table 7.6 Export Specialization According to Technology Classes in 1989 and 1994 – Share of Specific Product Group Exports in Total Exports of Manufactured Goods (in %)

	Czech Rep.		Hungary		Poland		Slovakia		Slovenia	
	'89	'94	'89	'94	'89	'94	'89	'94	'89	'94
Human Capital Intensive	n.a.	35.0	20.9	41.8	22.5	27.4	n.a.	25.5	n.a.	39.4
of which										
high tech	*n.a.*	*4.1*	*4.5*	*7.0*	*1.7*	*3.7*	*n.a.*	*2.2*	*n.a.*	*4.0*
mid tech	*n.a.*	*25.3*	*13.7*	*23.1*	*13.5*	*18.5*	*n.a.*	*18.6*	*n.a.*	*30.4*
Physical Capital Intensive	n.a.	4.2	1.7	1.2	1.5	1.1	n.a.	3.4	n.a.	2.0
Labour Intensive	n.a.	34.4	26.2	39.3	21.3	41.1	n.a.	33.5	n.a.	38.6
Resource Intensive	n.a.	23.8	21.2	16.4	22.7	29.0	n.a.	36.4	n.a.	19.2
Others	n.a.	1.0	1.8	0.5	1.4	0.6	n.a.	0.6	n.a.	0.2
All Manu-factured Goods	100	100	100	100	100	100	100	100	100	100

Source: OECD 1998, p.61.

Note: 1989 figures for Czech Republic, Slovakia and Slovenia were not available, since these states were not independent entitities then.

Table 7.7 discusses the growth of the private sectors in the Czech Republic, Hungary, Poland and Slovakia with respect to the private share in GDP and employment. In terms of the extension of privatization Hungary was second to Poland in 1990, and still is second in 1997, this time to Slovakia. In general, the changes in ownership have been fastest in the agricultural and service sectors and slowest in industry.

Table 7.7   Private Sector Growth in Four CEFTA Countries, 1991 and 1997 (% of the Total)

	Czech Rep.		Hungary		Poland		Slovakia		Slovenia	
	'90	'97	'90	'97	'90	'97	'90	'97	'90	'97
GDP	17.3	74.4	18.0	75.0	42.1	65.0	25.0	82.6	15.7	50.0
Employ-ment	18.8	57.3	48.0	60.0	55.5	67.0	25.8	63.1	17.5	n.a.

Source: BMWi 1998, Table 1.1.

Note: GDP data for the Czech Republic are from 1996, not 1997; employment data for the Czech Republic and for Hungary are from 1995, not 1997.

Table 7.8 shows the moves of the consumer price index in the five countries from 1991–1997. The data indicate that Hungary deferred a number of stabilization measures. While in 1991 the country was envied for the low inflation rate, in fact the lowest of all the five national economies, five years later Hungary's economy had to bear the highest inflation rate of the CEFTA Five. The conservative Antall/Boross administrations, faced with an extensive social security net until 1994, attempted to soften the hardship on the population with a mixture of half-hearted austerity programs and the devaluation of the currency rather than by strict macroeconomic stabilization. The austerity plan of the left-liberal Horn administration, which was introduced in early 1995, changed this policy. In fact, the attempt to balance the macroeconomic data was largely successful, but it came at the price of further losses in real income of the majority of the population and a stagnating economy.

Table 7.8   Growth of the Consumer Price Index for the CEFTA Five, 1991–1997 (% Change)

Country	1991	1992	1993	1994	1995	1996	1997
Czech Rep.	56.7	11.1	20.8	10.0	9.1	8	8.5
Slovakia	61.2	10.0	23.2	13.4	9.9	5	6.1
Poland	71.1	43.0	35.3	32.2	27.8	20	15.9
Slovenia	117.7	201.3	32.3	19.8	12.6	9	9.1
Hungary	35.0	23.0	22.5	18.8	28.2	23	18.3

Source: Kopint Datorg, 'Economic Trends in Eastern Europe', Budapest, 3/1998, p.183.

The next table reflects the development of the unemployment rate in the CEFTA Five. The figures are another indication for the problems Hungary has been running into as an effect of belated stabilization measures. Leaving the exceptional case of the Czech Republic aside, Hungary had by far the lowest unemployment rate among the CEECs in 1991. While this was still the case in 1997, the distance between Hungary and Poland, featuring the next lowest unemployment figures, has diminished. More importantly, in 1996 Hungary was the only economy with growing unemployment figures among the CEFTA Five (still leaving aside the atypical Czech Republic).

Table 7.9   Unemployment Rate for the CEFTA Five, 1991–1997 (% Rate Change to the Active Population at the End of the Year)

Country	1991	1992	1993	1994	1995	1996	1997
Czech Rep.	4.1	2.6	3.5	3.2	2.8	3.5	5.2
Slovakia	11.8	10.4	14.4	14.8	13.1	12.8	12.5
Poland	11.8	14.3	16.4	16.0	14.9	13.2	10.5
Slovenia	10.1	13.4	15.4	14.2	14.5	14.4	14.8
Hungary	7.5	12.3	12.1	10.4	10.4	10.7	10.4

Source: Kopint Datorg 3/1998, p.183.

Finally, table 7.10 condenses the transition efforts of the CEFTA Five and Russia in three indices. The European Bank for Reconstruction and Development (EBRD) measures the progress of reform on a scale of 1 to 4 in the following nine areas: large and small scale privatization, enterprise restructuring, price liberalization, competition, liberalization of trade and foreign exchange systems, banking and other financial reform. Out of all these indices the Economist Intelligence Unit (EIU) built an aggregate index, comparing it with its own two indices ('Transition', July-August 1996, p.9). Hungary and the Czech Republic score highest in most indices. The only exception is Slovenia, which gets a slightly higher mark on the EIU's Index for Institutional Development than the two others, ranking, however, decisively lower on the EBRD's Index. To clearly emphasize the distance the CEFTA Five have in their pace of reforms to the other countries in transition from realsocialism to capitalism, Russia has been included in the table. Unsurprisingly, the country scores much lower than all the CEFTA Five in all categories.

Table 7.10   Indices of Progress in Institutional Reform in CEFTA Five and Russia, 1996

	EBRD Index	EIU Index Institutional Dev.	EIU Index Political Dev.
Czech Republic	3.3	3.3	3.5
Hungary	3.3	3.3	3.5
Poland	3.1	3.3	3.3
Slovakia	3.0	2.7	2.5
Slovenia	2.9	3.4	3.5
Russia	2.3	2.3	2.1

Source: 'Transition' July-August 1996, p.9.

Summing up, a comparison with the other CEFTA Five shows that Hungary has economically done well during its transition back to capitalism. Nevertheless, the deferred stabilization measures have delayed – but perhaps also facilitated – the strong and steady upswing of the Hungarian economy. The center-left Horn government has advanced strategies regularly proposed by followers of neoliberal strategies, without really applying a full-fledged 'shock-therapy'. As a result, the macroeconomic imbalances of the national economy have been reduced, unemployment is among the lowest and the privatization level is among the highest in the CEFTA Five. Moreover, as a result of the 8% import duty introduced in March 1995 – which has been abandoned since then – the trade balance has somewhat recovered.

Perhaps more importantly, Hungary has began to shift its export specialization towards more human capital and technology intensive goods as early as in the first years of the latest transition. In all likelihood this difference is largely due to the strong foreign direct investment. Of course there is a danger in this development: it is unclear how pronounced a possible 'dual economy' has become in Hungary. In the last years of the 20th century there seem to be profitable foreign-owned firms with international best practices, modern equipment and management techniques, primarily operating in the mid-technology range, and often barely profitable Hungarian-owned and operated firms with old equipment and management practices, primarily in the low-technology range coexisting side by side. A precondition for Hungary's ability to overcome the dual economy and for the indigeneous part of the economy to learn from the foreign dominated part, is the existence of a rich knowledge base in the country. The next section shall take a look at the Hungarian knowledge base by analyzing the in- and output factors of the S&T system.

7.2 Hungary's S&T System in the Region

This partial success of the Hungarian economic restructuring efforts during the latest transition in comparison to the other Central and Eastern European Countries (CEECs) can also be seen in the development of the S&T system during the same time span. While the losses in input factors, mainly in capital and in personnel, have been substantial for Hungary's S&T system, they have been in line with developments in the other Central and Eastern European Free Trade Association (CEFTA) Five, the Czech Republic, Hungary, Poland, Slovakia and Slovenia. Regarding the output factors, data suggest that Hungary's S&T system is on the top end within the CEFTA Five. The first set of data presented in table 7.11 reflects the R&D expenditure of the CEFTA Five from 1991 to 1995. The numbers are barely comparable; however, clearly negative trends in the expenditure of each country can be discerned. One might well surmise that if the R&D

expenditures were measured strictly according to the criteria of the Organisation for Economic Cooperation and Development (OECD), they should be below 1% of GDP for all of the CEFTA Five. Nevertheless, with regard to the situation in Hungary, the European Commission has pointed out that the country 'experienced the largest loss in R&D potential compared to the other [CEE; PB] nations, considered its R&D expenditure in 1995 was only 26% of that in 1980' (European Commission 1997, p.402).

Table 7.11   R&D Expenditure in the CEFTA Five, 1991–1995
(in Percentage of GDP)

	1991	1992	1993	1994	1995
Czech Rep.	2.12	1.83	1.35	1.25	1.15
Hungary	1.07	1.05	0.98	0.89	0.75
Poland	1.05	0.83	0.83	0.82	0.75
Slovakia	2.25	1.88	1.53	1.01	1.04
Slovenia	n.a.	n.a.	1.77	1.46	n.a.

Sources: Czech Republic: OECD 1998, p.16; (note: until 1994 not entirely comparable to OECD standards; break in methodology with 1995); Hungary: OECD 1998, p.16; (note: until 1993 according to UNESCO standards; switch to OECD standards in 1994); Poland: OECD 1996, p.193; data for 1995 from OECD 1998, p.16; (note: until 1993 capital expenditure in enterprises and the higher education sector not included, depreciation costs included); Slovakia: OECD 1996, p.193; (note: until 1993 total expenditure of the R&D base, depreciation costs included); Slovenia: Statistical Office of the Republic of Slovenia, 11–03–96; (note: figures are overstated, as business expenditures most likely include non-R&D activities; for 1991 and 1992 the publications of the Statistical Office included only governmental R&D expenditures).

The trends in the numbers of R&D personnel working in the CEFTA Five are a reflection of the sinking rates of expenditure on R&D. In most countries the numbers of researchers have been falling dramatically. Again, comparisons between the countries are barely possible, as figures are notoriously unreliable, but trends can be discerned.

Table 7.12   Total R&D Personnel in the CEFTA Five 1991–1995 (in Full Time Equivalents)

	1991	1992	1993	1994	1995
Czech Rep.	81.896	60.292	40.793	37.779	22.678
Hungary	29.397	24.192	22.609	22.008	19.585
Poland	n.a.	n.a.	n.a.	79.248	83.590
Slovakia	12.576	10.681	8.927	10.249	16.183
Slovenia	6.990	7.031	6.143	5.841	n.a.

Sources: Czech Republic: OECD 1998, p.17; (note: data until 1994 overestimated; break in methodology in 1995); Hungary: OECD 1998, p.17; Poland: OECD 1998, p.17; Slovakia: OECD 1996, p.193; European Commission 1997, S–25, (note: until 1993, average recalculated number of full time RSE and technicians; with the exception of 1995, the numbers are grossly underestimated); Slovenia: for 1991, 1992 and 1993: Statistical Office of the Republic of Slovenia 1995; for 1994: Statistical Office of the Republic of Slovenia 11–03–1996 (note: the numbers are underestimated).

Of similar interest as the input are the output factors of the CEFTA Fives' S&T systems. In 1997 the European Commission attempted an analysis of the world shares in patents by countries. In 1994 both Hungary and Poland featured a lower share of world patents than they did in 1990. Surprisingly, the data for the Czech Republic and Slovakia seemed to indicate a rise in patent activities, which however was not explained by the authors of the report.

Table 7.13   Country Share of World Patents after EPO and USPTO for 1990 and 1994

	1990 World Shares		1994 World Shares	
	EPO	USPTO	EPO	USPTO
Czech Republic and Slovakia	0.04	0.01	0.05	0.02
Hungary	0.15	0.07	0.12	0.02
Poland	0.04	0.01	0.02	0.00

Source: European Commission 1997, p.428.

Note: EPO stands for European Patent Office, USPTO for United States Patent Office.

The table above reflects the international patenting activities of four of the CEFTA Five in 1990 and 1994. With regards to the output indicators of the system, the European Commission in 1994 stated, 'the Hungarian S(&)T (S)ystem performed better than other economies in transition...in many respects' (European Commission 1994, p.189). The Commission specifically pointed out that the number of Hungarian research papers held steady, with its overall share declining, however. Moreover, Hungary was the only CEEC among the top 20 countries in terms of citations per scientific paper (European Commission 1994, pp.196). Three years later the situation changed only insofar, that Poland featured stronger growth in publication and citation numbers from 1993–1995 than did Hungary (European Commission 1997, S–49, S–51).

In 1995 an effort was undertaken by Tibor Braun and András Schubert from the Budapest-based Information Science and Scientometrics Research Unit (ISSRU) to analyze the 'Indicators of Research Output in the Sciences From 5 Central European Countries'. Four of the countries analyzed in the research project are part of the CEFTA Five, namely the Czech Republic, Hungary, Slovakia and Slovenia. These four and Austria were taken into account because they lie within countries are situated on the area formerly occupied by the Austro-Hungarian Empire.[6]

In the research project mentioned above, Austria, the Czech Republic, Hungary, Slovakia and Slovenia were compared with respect to their research output in the life sciences, physics, chemistry, engineering and mathematics. The data were drawn from the Science Citation Index (SCI) of the United States' Institute for Scientific Information (ISI). Articles, letters, notes, reviews and meeting abstracts published in media recorded by the SCI were primarily included in the analysis.

Table 7.14 compares the publication output of the five countries. Austria has the largest number of publications in absolute terms as well as relative to the number of inhabitants. With respect to the number of publications per inhabitant, the other four countries are quite close to each other, with Slovenia retaining a small lead over the Czech Republic, Hungary and Slovakia. The following table reflects the absolute output figures of the five countries in all five fields combined for the period of 1990–1994 as well as the average number of yearly publications per 1000 inhabitants.

Table 7.14 Total Number of Publications per Country, 1990–1994
Average Number of Yearly Publications per 1000
Inhabitants (in Brackets)

Country	Austria	Czech Rep.	Hungary	Slovakia	Slovenia
Number of Publications	20.322 (0.511)	13.103 (0.252)	11.644 (0.226)	5.974 (0.221)	2.521 (0.256)

Source: Publication Figures from Braun/Schubert 1996; Population Figures from the World Almanac and Book of Facts 1995; own calculations.

The next table presents the percentage of publications in English. These figures could serve as a first indicator for the level of internationalization of the respective countries' national research efforts. An exception might be Austria, whose lagging behind might be explained by the fact that German is the secondmost important publication language in Central Europe, utilized more by the Austrian researchers, however, than by researchers from the other four countries in the sample.

Table 7.15 Share of English as Publication Language, 1990–1994

Country	Austria	Czech Rep.	Hungary	Slovakia	Slovenia
Publications in English	85%	94%	95%	92%	99%

Source: Braun/Schubert 1996.

Table 7.16 lists the six countries with which the highest number of cooperative publications have been produced for each country. In such a ranking of cooperation partners Germany has a small lead over the United States, both of which are ranked as number one partner for two countries and number two for two other countries. It is rather surprising that Russia, retaining the core of the former Soviet Union's S&T system, is not involved more strongly in the S&T efforts of the CEECs. After all, the Czech Republic, Hungary and Slovakia had been partners in the COMECON, with decade-old strong ties to the former Soviet Union.

Perhaps even more surprisingly, regional cooperation plays a small role in the S&T cooperation activities of the four CEECs currently in transition. In all four countries cooperation with large affluent Western societies is stronger than with the small neighboring countries. An exception here are Slovakia and – to an extent – the Czech Republic, which cooperate with each other, as well as Slovenia, which works together with Yugoslavia in S&T. However, it should be noted that in all these cases the involved partners have been in a political union during part of the time period analyzed (Czechoslovakia and Yugoslavia). With the exception of a Swedish-Slovene axis, all the other four CEECs currently in transition have no S&T cooperation with small countries. The picture for Austria is different, as the country has intense S&T links to all Western neighbors, including small Switzerland. The weak regional cooperation in S&T might, yet again, be explainable with the help of the center-periphery model as depicted in the previous section of this chapter. The model predicts that one finds a strong collaboration of the small CEECs in the economic periphery of Europe with the countries in the economic center of the continent, but only weak collaboration amongst the countries within the economic periphery. As table 7.16 shows, exactly this seems to be the case.

Table 7.16   The Six Most Frequent Cooperating Partner Countries and the Respective Number of Cooperative Articles, 1990–1994

Austria	Czech Rep.	Hungary	Slovakia	Slovenia
Germany (2423)	Germany (1130)	USA (1660)	Czech Rep. (402)	USA (288)
USA (1633)	USSR (878)	Germany (1197)	USSR (366)	Germany (253)
UK (642)	USA (677)	France (479)	Germany (310)	Yugoslavia (225)
Switzerland (641)	France (475)	UK (468)	USA (234)	UK (134)
France (610)	UK (430)	USSR (444)	France (158)	Sweden (129)
Italy (486)	Slovakia (402)	Italy (374)	UK (124)	Italy (125)

Source: Braun/Schubert 1996.

A comparison of the importance of the four fields of research in the five countries is highly interesting. The Austrian publications are, rather typically for Western capitalist systems, dominated by the life sciences. While the life sciences are responsible for almost two thirds of the country's publication output, the shares of the other sciences in the publication output of the national S&T system are smaller in Austria than in any other country in the sample.

The structure of Hungarian research is closest to the Austrian one. Here the life sciences are relatively more important in comparison to the other fields than for the other three CEECs, just as physics and engineering are relatively less important. An exception to this is mathematics, which has historically been a stronghold of Hungarian science.

In Slovenia chemistry is relatively less dominant than in the other three CEECs. The importance of physics and engineering relative to the other research fields is not surprising in this respect, given the dominance of the large Jozef Stefan Institute, which has specialized in physics, over the country's other research establishments. Remarkable, however, is that only in Slovakia the life sciences are of less relative importance than in Slovenia. Given the specialization and quality of Slovenia's medical research, this figure would need a more detailed analysis.

As might have been expected, the figures for the Czech Republic and Slovakia are similar to each other. This is not very surprising, given that both countries were part of the CSFR until 1993, more than half of the time span analyzed. In both countries chemistry is relatively more important than the other research fields.

In general, the research sphere of the CEECs are relatively more dominated by the 'hard' natural sciences, physics, chemistry and engineering, than is the case in Austria (and most other Western countries). These differences can be explained by the structure of the S&T systems of the formerly centrally planned economies, in which, as has been pointed out repeatedly, the natural and technical sciences were dominating over life and social sciences as well as humanities. A certain exception to this is the Slovene S&T system, which neither featured a large Academy of Sciences nor universities with the concomitant rigid separation of research and teaching functions. Nevertheless, in Slovenia realsocialism asked for a strong research effort in the natural sciences, too.

Table 7.17   Share of Major Fields, 1990–1994

	Austria	Czech Rep.	Hungary	Slovakia	Slovenia
Life Sciences	61.85	37.85	44.12	36.73	37.84
Physics	19.42	24.91	22.01	25.26	30.20
Chemistry	11.26	28.84	26.24	29.59	18.59
Engineering	9.85	13.26	11.08	14.65	18.32
Mathematics	2.08	2.29	4.68	2.09	3.55

Source: Braun/Schubert 1996.

With respect to the publication output of the five countries in all five fields combined, table 7.18 displays the figures for each year between 1990 and 1994. During this time span Austria shows a persistent growth in the publication output of the five fields. Between 1990 and 1994 the Austrian output grew by 26%. The Czech output grew by a mere 7%, the Hungarian one by 12%, the Slovakian by a remarkable 53% and the Slovenian by a strong 44%. Put in perspective, since Slovakia has the lowest per capita output in scientific publications of the five countries in question, one can speak of a catching-up process of the country's S&T system rather than of an attempt to leapfrog the other countries. Taking into account the output per capita, Slovenia's growth rates are most astonishing, as the country ranks second behind Austria in the amount of scientific publications per capita, but despite this high level still features a strong growth rate. Part of an explanation might be the emphasis of the Slovene scientists on their publication record, which is used as a yardstick for the possibilities of professional advancement by the respective authorities in the country.

Table 7.18   Publication Output for All Five Fields Combined, 1990–1994

	Austria	Czech Rep.	Hungary	Slovakia	Slovenia
1990	3.284	2.510	2.169	893	413
1991	3.410	2.323	2.370	1.002	384
1992	3.803	2.516	2.476	1.263	498
1993	3.884	2.617	2.379	1.315	530
1994	4.128	2.697	2.432	1.370	590

Source: Braun/Schubert 1996.

In conclusion, it can be stated that the Hungarian publication output did not decline during the latest transition from realsocialism to capitalism, but that it did not advance decisively either. The mere output figures of Hungarian science are comparable to the figures of comparable CEECs. With respect to the quality of Hungarian research output, however, the higher rate of citations from Hungarian papers may indicate a higher level of scientific advancement of the national research effort during the first years of transition from realsocialism to capitalism.

Having taken a comparative look at the national economies and the S&T systems of a number of countries in the region, the next step in the evaluation of the Hungarian development during the latest transition will consist of an analysis of the Austrian, Slovene and Hungarian S&T systems. And again special emphasis shall be laid on the most recent developments of the three countries' S&T systems.

7.3 The S&T Systems of Austria, Slovenia and Hungary Compared

Structures and functions of S&T systems in the region of CEE vary to quite a degree. Here, a short comparison of the S&T systems of three countries shall be attempted, one country with a longer tradition in capitalism, namely Austria, and two countries in transition from realsocialism to capitalism since 1989, namely Slovenia and Hungary. In an effort to condense the quintessential structures and functions of the S&T systems, only the core areas of the respective systems will be described and compared.

The *Austrian S&T system* bears the imprints of the political system, which, by and large, is still dominated by the corporatist 'Sozialpartnerschaft' (social partnership). The 'Sozialpartnerschaft,' which is not only a form of organization, but has really been a guiding principle for Austria since WWII, has been established in order to guarantee a high level of cooperation and conflict-free decision-making amongst governments, unions and chambers, the latter two representing the interests of employees and employers. Without doubt, the 'Sozialpartnerschaft' was an important factor in the economic success of post-WWII Austria. However, the corporatist structure has proven to be less efficient when it comes to the need to react flexibly to new developments as, for example, the rapid globalization of formerly national economic structures. Equally important, the 'Sozialpartnerschaft' is a centralized form of organization, which has serious democratic deficiencies as it has never been fully formalized and regulated by law. Despite the overall changes in the Austrian political system during the last decade, the Austrian variant of corporatism still operates by predetermining decisions, which according to the Austrian constitution would have to be made in the two chambers of the National Assembly. Leaving the effects of the 'Sozialpartnerschaft' aside, democratic functions are lacking also in a second area of the political process: the central bureaucracy of the country has a sizable political influence via the preparation and implementation of policies originating in the Germanic tradition of highly specialized, centralized and meritocratic bureaucracies.[7]

Exactly the same attributes which can be found in the political system are significant for the S&T system. The Austrian S&T system can be described as centralized, undemocratic, rigid in respect to new external developments and until the 1980s featuring a high level of conflict-free decision making.[8] Despite the raising of the voices of the S&T systems' different interest groups,[9] externally and internally triggered changes have been minimal so far. Most reform implementations were regularly planned and carried out by the central bureaucracy.

In fact, the structures of the 'Sozialpartnerschaft' can also be found in the S&T system of the country. In 1972 the 'Bundesministerium für Wissenschaft und Forschung' (BMWF, Federal Ministry for Science and Research) was reformed and, amongst other changes, given an internal structure of councils and coordination teams, which incorporate unions and chambers. In 1981 the 'Rat für Wissenschaft und Forschung' (Council for Science and Research) and the 'Konferenz für Wissenschaft und Forschung' (Conference for Science and Research) were founded as coordination and consultation fora for S&T, in which unions and chambers were included. In 1989 the 'Rat für Technologieentwicklung' (Council for Technological Development) was established, again with representatives of unions and chambers amongst its members (European Commission 1994, pp.196).

The influence of these corporatist institutions is limited, however, by a strong centralized ministerial bureaucracy, which before 1996 was part of three ministries. Then the 'Bundesministerium für öffentliche Wirtschaft und Verkehr' (BMöWV, Federal Ministry for the Public Sector and Traffic) was integrated into the 'Bundesministerium für Wissenschaft, Forschung und Kunst' (BMWFK, Federal Ministry for Science, Research and the

Arts), to form the new 'Bundesministerium für Wissenschaft, Verkehr und Kunst' (BMWVK, Federal Ministry for Science, Transport and the Arts), which, two years later, was turned into the 'Bundesministerium für Wissenschaft und Verkehr' (BMWV, Federal Ministry for Science and Transport). The BMWV exerts influence on the S&T system in that it creates guidelines, produces expertise for policy advice, formulates policies, is part of the decision-making process, and implements the policies once they have been agreed on.[10] Nevertheless, it is important to understand that in doing so the ministry arguably fills a void left open by politics, which sees S&T as a non-preferential policy field.

In fact, the National Assembly is comparatively uninfluentible with respect to the affairs of the Austrian S&T system. Only during the second half of the 1990s has a readiness to get involved into questions of S&T policy been signaled by the National Council, the more powerful of both chambers. The necessity of the National Assembly to get involved in S&T policy questions has become strikingly clear in the debates on genetic engineering and the EU-wide scandal around BSE, known also as 'mad cow disease'.

In figure 7.1, an attempt has been made to characterize the Austrian S&T system. To this end the technology system, the science system and the neighboring economic system have been arrayed on the x-axis. Furthermore, the respective systems have been analyzed according to two functional specializations, planning and acting functions, which have been lined up on the y-axis. Planning, or governance, functions can be found to be decreasing in ministries, agencies, funds, universities, and a number of other institutions, including the important intermediaries, which typically link scientific and industrial institutions. Acting, or performance, functions have been found to be decreasing in a number of lower level institutions, including intermediaries, as well as universities, funds, agencies and ministries.

On the ministerial level, besides the BMWV, the 'Bundesministerium für Wirtschaftliche Angelegenheiten' (BMwA, Federal Ministry for Economic Affairs) is the second important player in the field of S&T. An interministerial committee aims at the coordination of the two ministries with rather limited success. The committee also makes an effort to harmonize the policies of the organizational substructures of respective ministries, which regularly display their tensions quite openly.

The 'Fonds zur Förderung der Forschung der Gewerblichen Wirtschaft' (FFF, short: Economic Research Fund) is also part of the economic system; its counterpart, the 'Fonds zur Förderung der Wissenschaftlichen Forschung' (FWF, short: Scientific Research Fund) is part of the science system. Both funds have the character of agencies as they have a high public profile, and are actively involved in policy advice.[11] Especially the FFF bears the marks of the 'Sozialpartnerschaft', since the social partners are members of the board. The 'Innovations- und Technologiefonds' (ITF) and the ERP have sometimes been used to fund the ailing state-owned industry, but have increasingly become active technology policy instruments during the 1990s.

Since the 1980s the universities have been seeking stronger cooperation with the economy. To this end, a number of intermediary institutions has been founded, such as the 'Innovationsagentur' (innovation agency), the applied science oriented 'Christian Doppler Institutes'[12] and other less influential institutions, which have not been included in figure 7.1. Moreover, since 1988 the universities have been given more possibilities to engage into cooperations with industry under the heading 'Teilrechtsfähigkeit', an initiative providing the possibilities for university personnel to engage into industrial R&D and also earn money with these activities.

The Austrian 'Academy of Sciences' is a learned society, which also features a number of research institutes and disperses funds for the fostering of young researchers. The 'Boltzmann Institutes' are a large number of institutes working on basic and applied sciences, which have an administrative umbrella organization in common.

An array of intermediary institutions, located between industry and academia, are to be found in the comparatively large Austrian Research Center 'Seibersdorf', a number of smaller technology parks and centers, primarily located in economically less developed areas, and in the 'Johanneum', to name the most important. The 'Johanneum' is a research organization, which is partly financed by the 'Land Steiermark' (the state of Styria) and works on research contracts for the state of Styria, the federal state (Austria) and also private industry.

Since Austria's joining the EU, the Schumpeterian winds of destruction have shaken the Austrian economy, and hence the S&T system. Amongst other things the discussion on the evaluation of academia has been restarted (Salomon 1996, p.5). A network of 'Fachhochschulen' has been created, educating mid-level engineers. In addition to this the Vranitzky administration has decided to funnel an additional billion 1 Schillings, or million 90 USD per year into S&T until the year 2000 (Steiner 1996, p.20; 1997, p.14). Moreover, a program was devised during 1997, which was to address the Austrian backwardness in a number of key technologies by restructuring the national S&T governance structure and raising additional funds for S&T. Unluckily, major parts of the initiative were not carried out, because the ministers heading the BMWV and the BMwA were not able to reach a compromise in the ensuing struggle for competencies. After a lengthy discussion process, the only major change visible at the end of the decade is the creation of a new fund for centers of excellency in which industry and academia shall be enticed to work together more closely.[13]

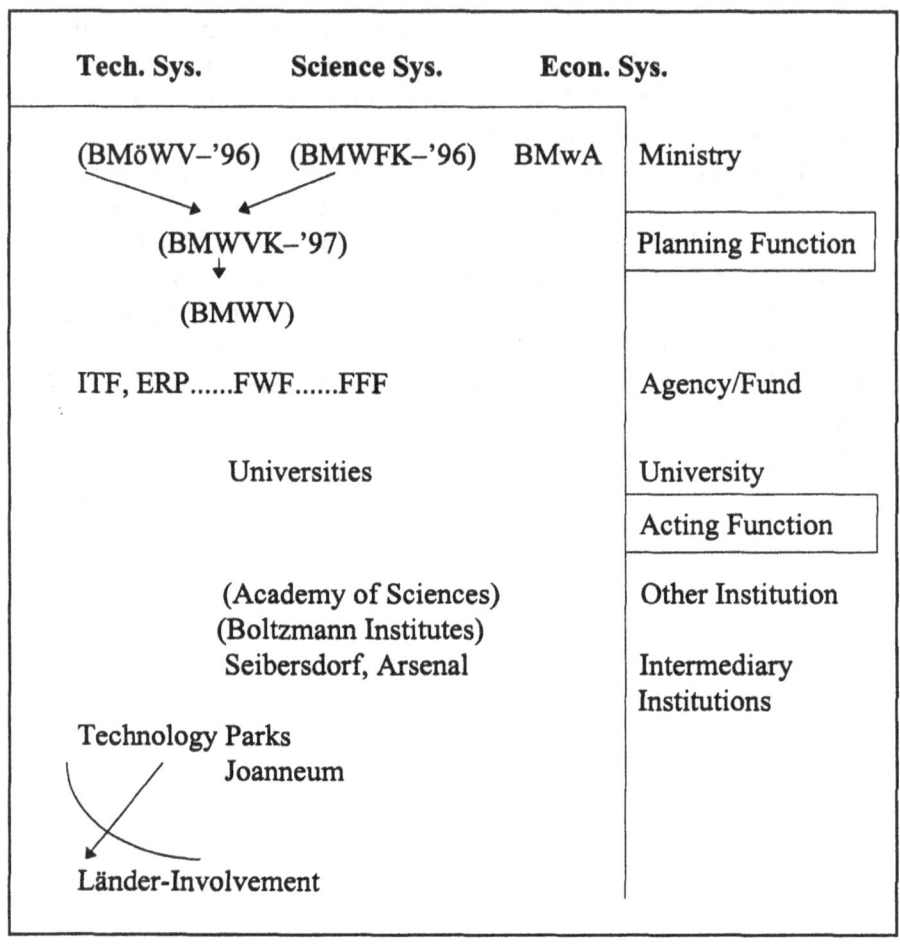

Figure 7.1   The Austrian S&T System

The *Slovene S&T system* is somewhat less centralized than its Austrian counterpart. A number of ministries are rivaling for a dominant role in the S&T system. Moreover, the efforts of several organizations, especially on the side of employers' organizations, to develop a corporatist structure similar to the Austrian 'Sozialpartnerschaft' have been stalled by the liberal center-right Drnovsek administrations. Consequently corporatist elements are not dominant in the S&T system, but can still be found in a number of councils, three of them inside the 'Ministry of Science and Technology' and one outside the ministry.

Nevertheless, in the daily policy process the primary financing of Slovene S&T lies in the responsibility of the 'Ministry of Science and Technology' (MZT). Insofar it is correct to say that the MZT is the most influential player in the arena of S&T policy. There are, however, a number of checks and balances which are to provide a sensible approach to policy making by the ministry.

Located inside the MZT is the 'National Council for Research and Development', the highest body of the ministry. Consisting of the chairpersons of the research councils which shall be described next and a number of experts, its primary task is to advise the minister on policy and research evaluation questions. In addition to this, the council suggests the distribution of funds among the programs of the MZT. The institution is an important player in the game for the internal budget distribution. Influential scientists find a forum to represent the interests of their respective disciplines in this body.

The six 'Research Councils' on natural, engineering, medical, agricultural, social sciences and the humanities are located a step closer to the researchers. Their task is to deliver information to the 'National Council for Research and Development' regarding their fields of research. Moreover, they assess projects and programs in their areas of expertise.

For the S&T system the 'Ministry for Education and Sports' is also of interest. The ministry plays a peculiar role in as far as it finances the wages and fixed costs of the tertiary educational system, including the university personnel, but has no policies evaluating, guiding or influencing the universities in any way. Not only does this give the universities much leeway, but it also leads to the question as to whether the funds are used efficiently or not. This very fact that the tertiary education system is financed both by the 'Ministry for Education and Sports and the Ministry for Science and Technology' has led to massive criticism, making the Ministry feel as if under constant siege. For this reason the Ministry seems to be in a position of permanent defense.[14]

Other ministries influence the S&T system, too, if in a lesser way. The 'Ministry of Finance' has a few regulations towards supporting investments into R&D by granting tax relieves. The 'Ministry for Economic Affairs' has a program to help SMEs to innovate.[15] Similarly the 'Ministry for Economic Relations and Development', the 'Ministry for Transport and Communications' and others have small policy programs influencing the S&T system, but without a larger coherent policy framework.

Standing between the ministries, the 'Committee for Research Coordination' is used as a communication platform. However, it seems to be difficult to achieve exactly the function suggested by the very name of the committee – as might be said about most similar institutions in most governments, including the Austrian one.

The Slovene 'National Assembly', the first chamber of the Parliament, and its 'Committee on S&T and Development', which plays an important role in questions of budgeting and legislation, are more directly involved into S&T questions than their Austrian counterparts. Similar to the Austrian second and less influential chamber, the Federal Chamber '(Bundesrat)', the Slovene 'National Council' is rarely concerned with questions of S&T.

An institution which has more influence in S&T policy making than its Austrian counterpart, the 'Council for Science and Research', is the 'Science and Technology Council of the Republic of Slovenia'. The body was established in 1992. Some of its members are appointed by government, some are members because of their functions (heads of the universities, the Academy of Sciences and the Chamber of Commerce). The council has a direct influence on the National Research Plan and other laws guiding the national S&T effort. It also evaluates the international competitiveness of S&T in Slovenia. Moreover, the institution has a strong informal influence on the National Assembly. It is the only advisory body in the S&T policy arena, which works at the level of government and not inside the Ministry of S&T (MZT).

According to the MZT, there are currently around 950 research groups in Slovenia. All of these are in some way affiliated to R&D institutions,[16] which can be divided into five groups:
– the two universities and other tertiary educational institutions,
– the nationalized research institutions,
– the independent research institutions,
– the Center for Scientific Research of the Academy of Sciences, and
– the business sector, consisting of commercial companies and public services.

The University of Ljubljana was formally established in 1919 and currently holds more than 25.500 students. The institution's 23 faculties, academies and colleges offer degrees in 248 fields. 1350 full and associate lecturers and 500 assistant lecturers are employed at the university, resulting in a student to teacher ratio superior to that of most universities in CEE.

The University of Maribor was formally established in 1975. It currently holds about 13.000 students, 220 full and associate lecturers and 120 assistant lecturers, which means that the student to teacher ratio is decisively less impressive than for the University of Ljubljana. Degrees in 131 programs are offered by the 8 faculties, schools and centers.

The two central national research institutions, comparable with the German 'Großforschungseinrichtungen', the Austrian 'Forschungszentrum Seibersdorf' or the United States' 'National Labs', are the 'Jozef Stefan Institute' and the 'National Institute of Chemistry'. With a total staff of 750 large by Slovene the 'Jozef Stefan Institute' is yet average by international standards. Its research fields are physics, chemistry, biochemistry, electronics, information science, energy research and nuclear technology. 81 scientists working at the institute have full or part-time responsibilities at the universities, which indicates some interaction between these institutions. With 170 employees, of which 76 were researchers in 1992, the 'National Institute of Chemistry' is already several categories smaller. Its main research programs are Structural and Theoretical Chemistry, Analytical Chemistry and Ecology, Organic and Inorganic Materials, Biotechnology and Chemical Engineering.

Other research institutions are nationalized, too: the 'Institute for Contemporary History', the 'Institute for Ethnic Studies' and the 'Urban Planning Institute' to give a few examples. The national institutes receive between 10–25% of their total budget as basic support from the state. In return, the state has a say in the management of the organizations, appoints the director and owns their property. The status of these institutes is unclear, since the basic funding is not enough to let them survive as a fully functioning organization. Several interview partners pointed out the unclear legal situation and the uncertain future of the institutes.

Of the somewhat larger number of independent research institutions, the Ministry of S&T listed 38 in its publication 'Science in Slovenia' (MZT 1994, pp.13–35). Most of them are located in Ljubljana, a few of them in Maribor. The organizations cover a wide variety of research fields and legal forms. Most of them are of small size. The majority of the institutes are located within the sphere of government so that the term 'independent institutes' probably refers to their independence from the university sector only.

Not only is the Academy of Sciences (SAZU) a society of distinguished scholars, as is the case in many countries in Western Europe and the US, but it also carries out research through an associated, but largely independent 'Center of Scientific Research' comprising 15 institutes. The main field of interest of these institutes, which are in a process of restructuring, is the study of Slovene culture and history. Contrary to the role of Academies of Sciences in other CEE countries, the SAZU seems to have no strong direct influence on S&T policies.

Moreover, the economy forms another sector of the R&D-performing institutions. Evidently, R&D in this sector has fallen dramatically, or, as one interview partner put it, 'instead of shedding fat, industry sheds brains'. While in comparison to other formerly real-socialist CEE countries the loss of R&D capacities seems to have been less painful, one has to take into account the small size of the national economy and the endangered 'critical mass' of R&D capacities in the country.

Last, but not least, there are 29 technology centers and parks, which are divided into three groups. Centers of applied knowledge are units within public research institutions or universities, which have separate accounting. Development units are organizations with separate accounting in industry. Technology parks and incubators ensure an environment in which start-up firms can develop (MZT 1994, p.9).

The largest technology park is located in Ljubljana, near the 'Institute Jozef Stefan' and a number of university departments and independent institutes. It has existed, if first under another name, since the beginning of 1993 and has seen a slow, but steady expansion since then. Co-founders include the 'Jozef Stefan Institute', the 'National Institutes of Biology and Chemistry', the 'Technological Development Fund', three large companies and SKB Bank. The foundation cooperates with the Ministry of S&T and the city of Ljubljana. In 1996 the technology park housed 12 companies with specializations ranging from measuring systems to computer applications and consultancies and the rationalization of combustion. In the same year three further proposals were analyzed by the park's management, with another eleven proposals in the making. The size of the park is physically and financially rather small and will most likely be too small to become an enterprise capable of covering its costs, despite its successes.

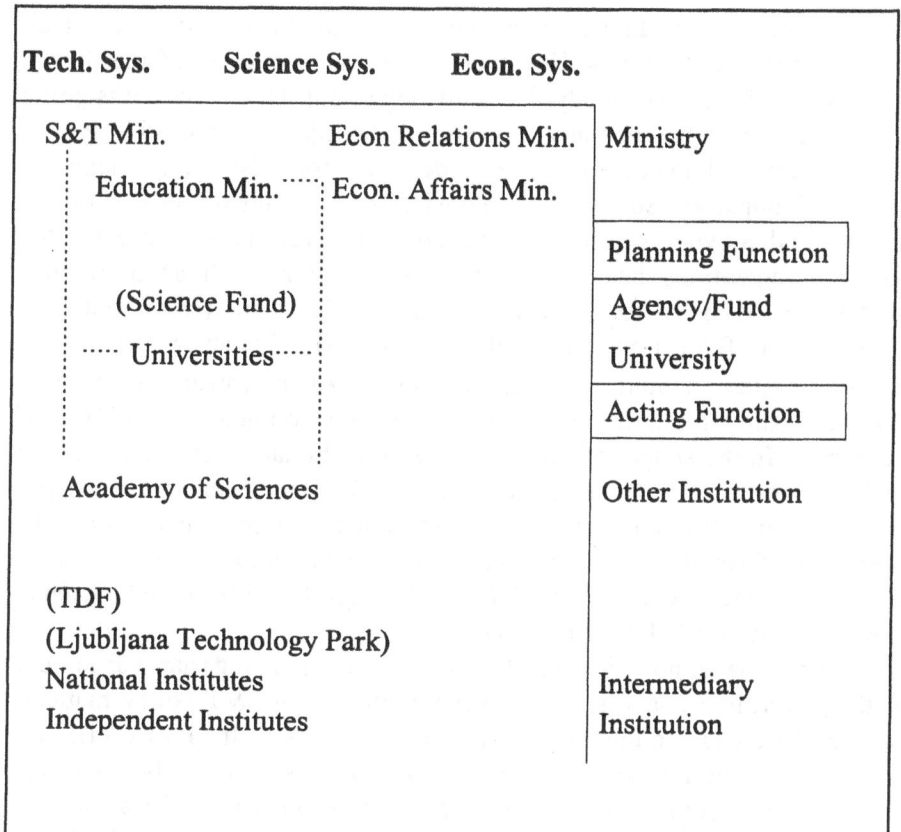

Figure 7.2 The Slovene S&T System

Since 1994 the *Hungarian S&T system*, as it has been described before and as it is depicted in figure 3, has gone through a cycle of (re-) centralization. With the reforms introduced by the Horn administration, the science and the technology ministries have been dissolved, the agency responsible for R&D (OMFB) has been scaled back in the bureaucratic hierarchy and the funds responsible for R&D sponsoring have been cut and reorganized as programs of the large 'Economic Development Fund' (GFA). A similar trend of centralization can be discerned in the higher education institutions, where scattered organizations have been reorganized since the early 1990s.

Contrary to the situation in Slovenia, the Horn administration does not oppose corporatist arrangements. Since 1996, the 'Chamber of Commerce' has been granted mandatory membership, one of the prerequisites of all strong corporatist structures. However, due to the weakness of the deeply split unions it is not very likely that the Hungarian political system is going to bear corporatist features any time soon. Consequently, the S&T system of the country sustains barely any reminders of corporatist arrangements.

The Hungarian S&T system consists of less institutions than the Austrian and Slovene counterparts do on each level. This is true for the highest organizations bearing governance functions such as ministries. Besides the 'Ministry for Industry, Trade and Tourism' (IKM) only the 'Ministry for Education and Culture' (MKM), financing the higher education system, should be mentioned here.[17] At the lower agency level only the 'National Committee for Technical Development' (OMFB) is of importance. In the sphere of funds or programs, the larger ones such as the 'Technological Development Fund/Program' (KMÜFA) and the 'National Research Fund/Program' (OTKA) are under direct supervision of government through the 'Economic Development Fund' (GFA), whereas the smaller ones such as the 'National High Priority Social Science Research Fund' (OKTK) are uninfluential.

Neither the 'Innovation Sub-Committee' of the Hungarian Parliament nor the government have shown too much interest in S&T policy matters. The Antall/Boross administrations deferred decisions on the Law on Higher Education and the Law on the Academy of Sciences until the last sessions of Parliament before the elections in spring of 1994. Drafts for a Law on Innovation were neither empowered by the Antall/Boross, nor the Horn administrations. Under the circumstances it was only to be expected that the highest institution advising government on matters of S&T, the 'Science Policy College', shared the fate of its predecessors in remaining largely uninfluential.

On the level of institutions primarily endowed with performance functions the situation is diverse, but not markedly different from the rest of the S&T system. In the higher education sector the drive to create associations has led to a number of such organizations, which include clusters of universities with total student numbers ranging from a few to up to twenty thousand. A number of critics have been calling for a further rationalization of the higher education sector (Bessenyei / Debreczeni / Setenyei 1994, Bessenyei / Melchior 1996). Yet, as a result of the austerity measures introduced by Minister of Finance Bokros in 1995, the financial situation of the universities is unsatisfying as most institutions are piling up debts presently. The government has promised to work out a plan of financial restructuring together with the universities and to raise the expenditures on higher education.

Contrary to the situation in the scattered higher education sector, the research institutes of the Hungarian 'Academy of Sciences' (MTA) are strictly organized under and controlled by their umbrella organization. In comparison to the Austrian and Slovene counterparts the MTA is endowed with much more leverage. While the latest transition to capitalism has reduced the number of the MTA's functions, the organization is still not only much larger, but also has a higher policy profile than the academies in the two neighboring countries, and still engages to a certain extent in S&T policy advice and making. Proof for the MTA's leverage has been the successful regaining of control over OTKA by the Academy, supposedly against the will of the 'Ministry for Culture and Education'.

Still one level lower, in the sphere of intermediary institutions, the most outstanding organization was also part of the innovations of the Hungarian S&T system during the 1990s: the 'Bay Zoltán Institutes'. The three institutes seemed to have had a good start, especially in light of the industrial R&D system's troublesome situation.

Among the other intermediary institutions S&T parks are currently being planned. Two large competing projects have been planned for Budapest. The smaller one, totalling USD 25 million, has been advanced by a private entrepreneur. The second project is almost 10 times as large, but has government backing (Strassel 1996, pp.18). Whatever project wins out, it will be the first sizeable technology park for Hungary. The chances for the success of such a project are not clear, since one the one hand the existing structures remind more of incubators and feature only low growth. On the other hand, the country has been experiencing the largest foreign direct investment inflow of all CEECs. If only a small part of this foreign capital can be diverted into a technology park, an interesting and potentially promising development could begin.

As a result of the austerity plan of the Horn administration and the general capital scarcity in the country, the Hungarian S&T system has been increasingly forced to look abroad for resources. As the EU has opened up its main R&D initiative, the Framework Programs, to the CEECs, Hungarian researchers are attempting to enter these funds. Specifically, the 'Bay Zoltán Institutes' are making an effort to get into the Framework Programs of the EU, as are the universities, which are also aiming at World

Bank loans. OMFB has put aside HUF 300 million for the co-funding of EU-research projects. With the 1998 decision to fully enter the Fifth Framework Programme, this sum will have to be substantially enlarged for the time span of 1999–2003.

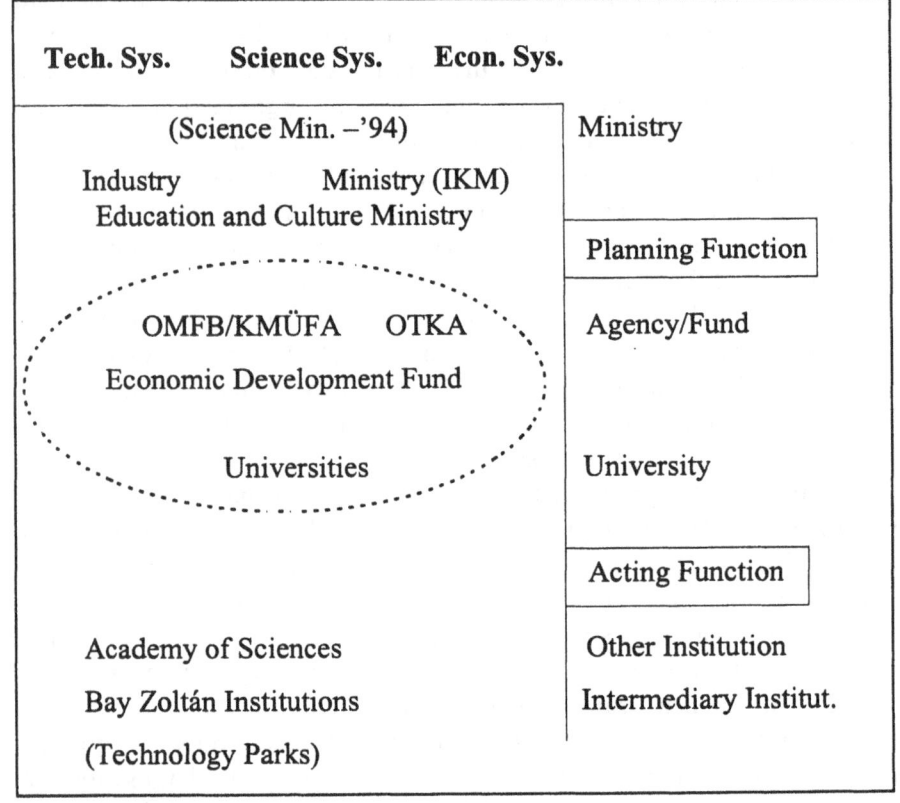

Figure 7.3   The Hungarian S&T System

As can be inferred from what has been said above, the S&T systems of the three neighboring countries Austria, Slovenia and Hungary differ in a number of respects. For instance, the degree of centralization is highest in Austria and lowest in Hungary, with Slovenia ranking in between. Austria also features the highest density of corporatist arrangements in its S&T system, again followed by Slovenia and Hungary. Regardless of the degrees of centralization and corporatism displayed by them, the role of a number of institutions differs in the three S&T systems.

Table 7.19 Influence of Institutions on the S&T System's Governance, Degree of Centralization, Density of Corporatist Arrangements, as of 1998

	Austria	Slovenia	Hungary
Influence of Institutions:			
Central Government	++	+	+++
Central Bureaucracy (Ministries)	+++	++	++
Agencies/Funds	++	–	+
National Assembly	+	++	+
Regional Governments	+	–	+
Intermediary Institutions	+	+++	+
Academy of Sciences	–	–	++
Universities	++	+++	+(+)
Large Enterprises	+	+	+(+)
Small and Medium Enterprises	–	–	–
Labor Unions	++	–	–
Chambers of Commerce	++	(+)	–
Aggregates:			
Degree of Centralization	+++	++	+
Density of Corporatist Arrangements	+++	+	–

Note: '+++' stands for strong/important, '++' for mediocre, '+' for low, and '–' for no influence/degree/density.

A number of patterns are recognizable from a comparison of the three national S&T systems. First, only two groups of institutions play an important role in all three systems: the central bureaucracy and the universities. The former group of institutions concentrates a number of functions in its hands, which clearly are not just simply planning-oriented, but rather aimed at a wider field of governance. Functions of the central ministries include policy preparation, formulation, accordation and implementation. The role of central ministries might be stronger for S&T policies than in other fields, because the S&T area is not regularly deemed as important as social or economic policy, on which the interest of the public is focused more or less constantly by policy makers. The latter group of institutions, which are playing an important role in all three national S&T systems, is that of the universities, which are the largest and strongest organizations, since it represent the interests of large numbers of scientists, an articulate and influential clientele. Due to the size and leverage of the Academy of Sciences this is true only to an extent in Hungary.

Second, in all three systems the National Assembly is far from being the most influential institution of the S&T system, a tendency which points at a severe democracy deficit in this policy field. This, again, might be due to the limited public interest in the S&T area. However, there are indications that this situation might change, especially in the more affluent Western European societies. Moreover, the complexity of the topics involved into S&T policy making is large, perhaps larger than in other policy fields. Finally, due to their education and social prestige, the scientists and engineers as the main clients of S&T policy makers are a highly unruleable group of people.

Third, the private sector has only little influence on the S&T systems. This is true both for private industry and private research institutions. Industry often has only short-term interests in S&T, i.e. firms want to reap the benefits of subsidies and other forms of funding, while preventing other companies from doing the same. Sometimes private research institutions have similar motivations, without being as established as their mostly larger competitors, the universities and public research institutions. As a result, both, private industry and research institutions have only a limited influence on S&T policy making.

Fourth, the three systems are dominated by different groups of institutions: Austria mainly by the central bureaucracy, Slovenia by universities and intermediary institutions and Hungary by the central government. In all three cases this is explicable by the specific historical circumstances leading to the present situation in the respective S&T systems. This observation might be interpreted as a high degree of path dependency for the development of structures and functions of the analyzed countries' S&T systems.

In Austria the central bureaucracy has been playing an important role in most policy areas since the days of the Austro-Hungarian Monarchy, when the center of administration of an empire of more than 40 million was Vienna. Moreover, a number of reasons mentioned before are responsible for the strong involvement of bureaucrats in the S&T governance, ranging from the complexity of the topics making them hardly understandable for policy-makers, the media, the public and even scientists of other disciplines, to the limited presence of politicians in this policy field as it is less likely to be the center of public discussion to such a degree as economic and social policies are.

In Slovenia the S&T system has a rather short history, since industrialization was belated due to the political constellations within the Austro-Hungarian Monarchy, the country was part of. Accordingly, the first Slovene university was founded only after WWI and the Academy of Sciences shortly before WWII. As a result, the structures of the S&T system were mainly created during realsocialism. There were specific differences between S&T governance in realsocialist Yugoslavia and the COMECON (Council for Mutual Economic Assistance) countries, which meant the absence of a powerful Academy of Sciences and a strong role for public research institutions as the Jozef Stefan Institute and the universities for Slovenia. Still, in Slovenia as in the COMECON countries, realsocialism was based upon Marxism-Leninism, which called for a strong role of the natural and technical sciences. Indeed, Slovene scientists had a strong voice during the times of realsocialism. The role of the scientists in the Slovene S&T system was balanced mainly by the corporations, the users of the knowledge generated. With the beginning of the transition to capitalism most large firms faltered or were severely downsized. From then on the counterweight to the scientists has been missing. As a result, the Slovene S&T system today is dominated by a single constituency: the scientists.[18]

In Hungary the Horn administrations have actively reformed the S&T system, which, moreover, was also affected by the Horn government's austerity package in 1995. Despite a wide-spread feeling amongst Hungarian analysts that the changes of the last years have had negative effects on the S&T system, it cannot be denied that the role of the Horn administrations in these changes has been proactive. Nevertheless, the Academy of Sciences, still bearing some of its leverage from the times of realsocialism, has a say in the development of policies. So do the universities, which profit from the relative loss of influence of the Academy. Large corporations play a somewhat more important role in the S&T system of the country, since the concentration of economic power in the hand of relatively few enterprises in Hungary is larger than in the other two countries. However, this influence is not directed at the core areas of the S&T system, but rather at questions of taxation, import restrictions and the like.

Notes

1   In February 1991 Czechoslovakia, Hungary and Poland formed the so-called Visegrád group signing a declaration in Visegrád, Hungary, which was based on agreements of economic, but also political cooperation. CEFTA was founded in December 1992 by the same countries, and was enlarged by Slovenia in January 1996 and by Romania in July 1997. Since then other countries have expressed their interest in joining CEFTA. By 1996 CEFTA had abolished duties on approximately 80% of industrial products and on a steadily growing percentage of agricultural products. (http://www.uvi.si/cefta/ at 23–02–1999).

2   Center-periphery models were developed during the 1970s, when they were primarily used to analyse the international political economy. Most of these models distinguished between national economies forming a center (usually the OECD countries), a semi-periphery (amongst others, CEECs, Southern Europe, the East Asian Newly Industrializing Countries) and a periphery (the developing countries). Center-periphery relations were described as dominated by the center, which would generate profits through importing raw materials and semi-finished products cheaply and exporting capital and consumer goods with large profit margins. Since in this model the center is economically much more powerful than the periphery, the terms of trade are dictated by the developed countries – much to the benefit of the developed world. In Europe the works of Dieter Senghaas and Johan Galtung on the subject have been very influential: see for example, Senghaas 1972, and Senghaas 1974. Internationally at least as influential was the Latin American strand of center-periphery models, 'dependencia' theories: see for example Cardoso 1971, or Galeano 1973.

3   An exception is the Czech Republic, which began to slide in a structural crisis.

4   In itself, this indicator is not enough to qualify the country as economically highest developed in the region. However, the figures are an indicator for economic development.

5   Human capital is another term for knowledge embodied in personnel, acquired through both formal education and training on the job.

6   If not otherwise stated the data presented in the rest of this section are drawn from Braun/Schubert 1996.

7   The Germanic countries' tradition of a Weberian meritocratic and independent, but heavy handedly reigning bureaucracy is illustrated by the Austrian saying, 'Regierung vergeht, Verwaltung besteht'. Freely translated this means, 'governments come and go, but the bureaucracy is here to stay'. In fact, in all the political turmoil between 1918, the breakup of the Empire, 1934, the Civil War, 1938, the annexation to Nazi Germany and 1945, the liberation by the Allied Forces, the Austrian bureaucracy was largely unchanged, always serving the respective political master. More interestingly, this behaviour is so intensively culturally ingrained that it has largely remained unchallenged by society – one might even say that this behaviour was expected by large parts of society.

8   Compare with Aichholzer/Martinsen/Melchior 1994, esp. pp.4.

9   While large scale protests at universities were largely organized by students in 1987 and 1991, in 1996, for the first time in recent Austrian history, both professors and students went out on the street for weeks and protested against the 'Sparpaket', an austerity package of the Vranitzky administration in the wake of the meeting of the EU's Maastricht criteria, and for reforms in the organization of science.

10   The leverage of the ministry has been criticized, amongst others, by the OECD team, which analyzed the Austrian S&T system in 1988. See OECD 1988. In early 1997, protests of the humanities faculties against the draft 'Studiengesetz', the law regulating curricula and the organization of the disciplines, were directed not against the minister, but directly at Sigurd Höllinger, a high-ranking bureaucrat responsible for the draft ('Der Standard', 16–01–1996).

11   For an analysis of and a number of reform suggestions for the two funds, see Campbell/Felderer 1994, pp.218.

12   For a description and analysis of the Christian Doppler Institutes, see Littig 1996.

13   See for example, 'Ressort für Wissenschaft, Forschung, Technologie', in: 'Der Standard', 14–03–1997, p.26; 'Die Technologiekonzepte stehen, aber Finanzierung noch offen', in: 'Der Standard', 26–11–1997, p.23 and 'Aus einem Büro wird nur ein Vorzimmer', in: 'Der Standard', 15–03–1998, p.14.

14   I have tried to contact the organization, but despite dozens of phone calls it was not possible to meet any higher ranking personnel from the Ministry for Education and Sports. Appointments would regularly be cancelled two or three hours before the meeting would take place.

15   However, a high ranking official of the Ministry has assured the author that not only has the ministry has no technology policy, but that he also thinks that it is wrong to have such a policy at all.

16   It should be mentioned that of course there are a number of researchers, who are unaffiliated. However, the number of these persons is rather small and is estimated at between 70–100 people.

17   For more information, see for example, Bessenyei/Debreczeni/Setenyi 1994, pp.37.

18   For a more elaborated analysis of the internal structure of the Slovene S&T system, see Biegelbauer 1996.

8 Learning From the Past for the Future

The final chapter consists of two sections. Section one evaluates these research questions and hypotheses which were laid out in the introduction. Section two draws a few general policy lessons from the analysis of the Hungarian S&T policies since industrialization.

8.1 The Introduction Revisited

Having reached the end of this study, it is time to revisit the research questions from the introductory chapter. One of the goals of the book was to learn more about the intertwined nature of the different forms of innovation, in science, technology, economy and politics. The second research question was directed at the nature and origin of those ideas and paradigmatic notions underlying the policies directed at the S&T system, which have been termed 'policy paradigms'. The third goal of this study was the evaluation of the S&T policies, with an attempt at explaining the successes and failures of the respective policies. For all three purposes the evolution of the Hungarian S&T system since industrialization was analyzed in a comparative perspective. During the course of this section these three research questions shall be reexamined.

With regards to the first goal, after conceptualizing society as consisting of a number of (subs)systems, it became obvious that – as was expected – science, technology, economic and political system all interact with each other. The actual process of this interaction is indeed problematic, since the internal logic of the four systems are far from being uniform.[1] The driving force for actors in the political system is to obtain power and to govern. This fundamental logic has not changed, regardless of the rapidly changing forms of governance to be found in Hungary since industrialization. Neither has the internal logic of the science system – the search for truth or a similar construct – nor of the technological system – the search for artifacts fulfilling human needs – changed during the time

187

period analyzed. The only system subject to fundamental changes in this sense was the economy: during capitalism the economic system featured as its main driving force the seeking of profits, whereas during realsocialism the socialization of the factors of production made this goal difficult, if not impossible; therefore the goal was supplanted by the optimal allocation of goods according to societal needs, a more bureaucracy-steered and group-oriented aim, which led to a number of problems in the economy and was arguably a critical factor for the demise of realsocialism.

What did change over time, however, was the framework setting the parameters on the bases of which these logics could operate. During the monarchy, the power of the Hungarian aristocracy found only little counterbalance in other endogenous societal interests – a fact subject to change only with the rise of industry, which produced a small bourgeoisie in Hungary. Later, first during authoritarian and then fascist regimes, the discretionary power of the politicians expanded again, yet it was only during Stalinism that the possibilities of the Communist elite to rule the country through a centralized command structure reached its maximum. Since then, as an effect of rising pluralism first in the party structures, then throughout society, the steering power at the discretion of politicians has been diminishing.

These changes in the parameters of the political system had direct effects on the framework under which the economic system operated. The economy from the early 1940s to the late 1980s became subjugated under the control of the political system, first as an effect of the fascist war economy, then of realsocialist strategies. The latest transition, from realsocialism to capitalism, changed this situation and lessened the controls of the political system over the economy considerably.

With the notable exception of the interwar period, science and technology took an upswing with regards to their material basis and their societal status, from the 1870s until the beginning of the latest transition in the 1980s. As has been pointed out, science was even declared a 'direct force of production' by the Communist party propaganda. Not only did the function of science change with the latest transition, but also its public image, which seemed to decline, as did its level of resource endowment, which sank considerably.

But, returning to the question posed before as to the nature of the relationship between science, technology, economy and politics – did the relationship itself change fundamentally during the time period analyzed?

The answer has to be split: on the one hand, each time there was a regime change, which in itself often came hand in hand with important changes in both world politics and the world economy, the position of the four systems relative to each other did indeed change. First of all, most of the time politics dominated the other systems; this seems to be less the case since the beginning of the latest transition – most clearly with regards to the economy. Another example is the science system's relative importance: in the time of realsocialism scientists seemed to have better and more extensive access to politics than during the 1990s – a fact, which might contribute to the explanation of diminishing resources for S&T since the beginning of the latest transition.

On the other hand, however, the basic quality of the four systems' relations to each other barely changed at all. As has been said, this relationship seems to be contingent on the internal logic of the respective systems. Apart from this observation, the only slowly changing relations between the four systems can be interpreted as an indicator for a general path dependency of the scientific, technological, economic and political systems. In fact, all four systems consist of a number of institutions, which themselves were subject to pronounced changes during each transition and even to several reforms in between. Yet the four systems retained a number of (core) institutions, which might have changed in appearance and name, but in their essence were permanent features of the respective systems. This is the case for the political system, which since the 1870s has featured a specific set of ministries with more or less unchanged functions – and which, at least during the latest two transition periods, seemed barely to exchange staff for political reasons, even in times of serious reform efforts.

This is also the case for the economic system, featuring an industrial structure, which transformed only quite slowly. Moreover, the relative importance of the respective economic sectors changed only gradually and over decades – regardless of regime changes and ideological differences between regimes, which found their reflection in often strongly differing policy measures. Examples are the rather procurement-oriented economic policies of the monarchy, the more direct dirigiste policies of the war economy and the time of massive interference in the economy by the state during Stalinism, which all led to a growing orientation of the economy towards heavy industries, which, however, was in line with the development of other national economies in the region. Indeed, the economic specialization of Hungary in relation to the respective structures of Austria and Germany remained stable over 130 years: all three countries first moved into light, then rapidly into heavy industries and electrical machinery and only quite recently shifted over to service industries. During this transformation the wealth creation of the respective economic sectors of the three countries, as expressed in GDP shares, did not change relative to each other – to stress this point again: this was the case, despite the Hungarian efforts during realsocialism and here especially Stalinism to massively invest into heavy industry.

Similarly, the specialization of Hungarian science in a number of fields such as mathematics or biochemistry was only influenced to a limited extent by the decision of the Communist regime to invest into technical sciences such as mechanical engineering and 'hard' natural sciences such as physics and chemistry. This does not mean that the steering efforts of realsocialist politicians in these respects had no effect at all, but rather that the actual changes they were able to produce were strongly contingent on the historical evolution of the Hungarian S&T system. This was an important reason as to why, despite politicians' efforts comparable to those in other Council for Mutual Economic Assistance (COMECON) countries, Hungary's S&T system was and is comparatively closer to a 'Western' set-up. This means that the technical sciences, although they are better represented in the Hungarian S&T system than in most West European and North American countries, are less well established in Hungary than in most other former COMECON countries. In turn, a classical strength of the Hungarian S&T system, mathematics, is stronger than in many former COMECON countries, as are the social sciences.

What can therefore be found is a surprisingly strong inertia of the intra-systemic structures of science, technology, economy and politics during transitions. The substantial degree of the four systems' resistance to change is indicated by the mostly unchanged internal logic and the persistent existence of their main institutions over time (if sometimes under different names and with slightly varying functions). These systemic rigidities are all the more impressive, when one considers that the formal set-up of the four systems, if measured by indicators such as official mission statements, organigrams or taxation structures, has sometimes been changed dramatically during the three transitions discussed in the study.

But how can it be that regime changes from fascism to Stalinism or from realsocialism to capitalism have only a comparatively small impact on the fundamental qualities of the science system or the relationship between the political and the science system?

One possible answer to this question is provided by the concept of 'social systems of production', which was outlined and even tested in the beginning of this book. One of the central propositions of the 'social systems of production' notion is that institutions of a social system of production are arranged in accordance with, and even through, ordering principles, such as the market, the hierarchy, or the network. These principles can be interpreted as a 'encoding mechanism' of a society's basic norms and values, which transforms the latter into institutions that can be understood as 'genetic code structures' of a higher order. As the institutions of a social system of production are embedded in and contingent upon a specific set of slowly changing norms and values, the concept predicts that social systems of production, in general, can change only gradually (compare with Boyer/Hollingsworth 1997, Hollingsworth 1998). Moreover, the model indicates that even if an institution tries to change its structures and functions, it is limited in its efforts to its linkages to other institutions of the respective social system of production, which also are bound to the very set of norms and value, which the institution dedicated to change, wants to overcome (see chapter 2.2).

These predictions in fact are in line with the observations made in the case of Hungary's scientific, technological, economic and political systems. Each time efforts were undertaken to change either singular institutions or the whole institutional set-up of Hungary, the effects were limited in comparison to the expectations of decision-makers. In a number of cases, some of which shall be discussed later on, even counter-productive results were realized.

Another explanation for the discrepancy between the seemingly fundamental nature of the reforms and the actual changes in the most basic workings of the societal (sub)systems might be provided by the second major research question posed in the introduction: what is the nature of the notions and ideas, which underlie policies, the 'policy paradigms'. It was shown that the paradigmatic notions underlying the Hungarian S&T policies have not necessarily changed in accordance with the time-frames of the historical transitions the country has gone through since industrialization. As these transitions frequently went hand in hand with changes in ideology, policy paradigms appear to be independent of specific ideologies. Since the policy paradigms, which by definition form the framework of decision-making for (here S&T) policies, have been found to change slowly and not in time with transitions, they can be seen as another explanation for the resistance to change in Hungarian S&T institutions.

Although the monarchical feudal-capitalistic economy, the fascist war economy, and the Stalinist centrally planned economy had a number of differences with respect to their structures and political means, their goals were the same, namely catching up with Western European economies. Moreover, during these time periods politicians' understanding of how to instrumentalize S&T for this goal was quite similar. They all implicitly based their S&T policies, educational policies and industrial policies on the linear model of technological change. Because of this, the S&T policies then became part of the corresponding 'science push' policy paradigm, which prescribed a strong S&T base as one of the necessary ingredients for a successful catch-up process. Similarly, the equally linear 'demand-pull' paradigm was underlying the S&T policies of both reformsocialist and the first post-realsocialist governments. These findings are an indication of the mechanisms responsible for the changes in Hungarian S&T policy paradigms, which not only seem to be largely independent of transitions, but indeed may be exogenous.

In fact, it has been shown during the course of the study that Hungary has taken models of technological change, S&T paradigms and the policies themselves from hegemonic powers. All three models and associated policy paradigms, the 'science push', the 'demand pull' and the 'innovation process' paradigm, have been taken from major powers – first Austria and Germany, then the Soviet Union, later the United States and, finally, Germany. Despite the fact that the paradigms were identical and the policies were quite similar to the ones of the respective major powers, cultural and historical factors specific to Hungary and Central and East Europe (CEE) have caused the actual outcomes of these S&T policies, to be unmistakably and indeed quite different from the outcomes of the policies of the major powers.

A number of examples described in the study come to mind: during Stalinism the scientific personnel of the universities was purged for political reasons – a transfer of Stalinist policy. However, a number of those scientists, deemed to be politically unreliable, were not restricted from the possibility of engaging in research, rather they were employed in the Academy of Sciences. Incidentally, this adaptation or variation of Stalinist policy had long-term effects: some of these 'hidden' scientists were engaged in the political uprising of 1956 and others took important political positions after the beginning of the latest transition from realsocialism to capitalism.

Another example is the introduction of demand-led policies as a result of the 1960s reform program, the New Economic Mechanism (NEM), which was an emulation of one of the basic mechanisms of the S&T system of the United States . The S&T system of the United States conforms to one of the dominant ordering principles of the country's social system of production, the market. The introduction of the NEM's S&T policies should have led business to receive some of the Hungarian scientists' know-how by contracting out R&D projects to scientists. Actually – and quite different from the situation in the United States – the measures led to a network of scientists employing each other in various ways and only slightly to the fostering of innovative activity in industry.

In addition to cultural and historical conditions, a number of specific circumstances had a major impact on the actual S&T policies. For example during the NEM Hungarian policy makers devised S&T policies according to the 'demand pull' policy paradigm, under the assumption that the linear model of technological change would explain innovative behavior. Despite this common conceptualization of the innovation process from the late 1960s to the early 1990s, expenditures on S&T rose until the late 1970s, then stagnated and finally diminished dramatically with the first post-realsocialist governments. Two factors explain this phenomenon. First is the Hungarian macroeconomic situation, which was marked in the second half of the 1970s by low growth rates and expansionist policies. These policies provided funds for state activities through foreign debt accumulation and increased budgetary pressures markedly. Second were the various interest groups within the S&T system, which found it hard to compete for funds with other societal subsystems. Despite the fact that the notion of policy paradigms adds an important dimension to the understanding of (S&T) policies, the concept that paradigmatic notions underlie policies cannot explain the actual policies without further qualification. In this case the explanation comes from the analysis of the circumstances and societal arrangements, where the policies were devised and deployed.

During the course of this study it has been shown that the four S&T policy paradigms, which have been employed in Hungary since industrialization, have been reactions to four distinct periods in the development of Hungary. During the first period, from the late 1860s to the early 1940s, the Hungarian state was formed, independence was achieved and both capitalism and fascism in their specific Hungarian variants developed. During this time, the prime role of S&T policy was to strengthen the 'nation' and to serve as a source of national pride. Technology policy, and to a certain extent also science policy, was not consistent during this period. The era of WWII was a watershed in the transformation of S&T policy, as it played a key role in the mobilization and build-up for the war effort. In the Cold War environment of the 1950s and 1960s, S&T policy was seen as a motor of progress and as a possibility to enhance military capabilities – although the Hungarian efforts to enhance the military capacities of the COMECON were limited. In both the Second World War and the Cold War the concept of technological change was based on the linear model, which came with the science push policy paradigm.

As an effect of the intermittent weakening of bipolarism, the wars in Vietnam and in Afghanistan, the oil and debt crises and the gradual rise of Japan and Germany, the science push policy paradigm during the 1970s was regarded as insufficient. S&T were seen increasingly as being accountable to society and were understood as a solution to various social, but also economic problems. In this respect environmentalism was important in many West European countries and in the United States, but became a political factor in Hungary only in the 1980s. The dominant views of technological change still were based on the linear model, but the policy paradigm underlying the actual S&T policies changed and can be referred to as the demand pull paradigm.

During the last decade of the 20th century, the international political economy has changed dramatically. The strengthening of the trilateralism of the US, the EU and Japan, the fall of the iron curtain, the intensification of international economic competition, rising unemployment levels in Europe and widening income disparities in most advanced industrialized societies, especially in the United States and the UK are evidence of this change. In this context S&T are increasingly seen as a source of strategic opportunity for national and regional economies. In the short term they enhance innovative activity and in the long term they lead to a more qualified work force.

With regards to the concept of technological change, the linear model still is important, although the more complex holistic model has been recognized in many instances as accounting more for real life phenomena. It is important to notice that in the advanced industrial societies many policies of the 1990s are not only based on conceptions of technological change following from the linear model, but that they are straight-forward applications of the demand-pull, and sometimes even the science-push policy paradigms. This seems to coincide with path dependencies of institutional arrangements around key technologies such as atomic power. These institutional arrangements were created in the 1950s and 1960s under the then dominant science-push policy paradigm and today still carry the conceptions ingrained in the institutional set-ups and finance mechanisms, which were constructed with policies based on the science-push paradigm (Stucke 1993, Grande/Häusler 1994). At the end of the 1990s, most S&T policies in Hungary are still based on the science-push and demand-pull policy paradigms.

Still, a noticeable change has taken place from these two policy paradigms to the increasingly important innovation process paradigm, as the latest paradigmatic notion underlying S&T policies is called here. While the exact features of the latest policy paradigm are yet to be determined, it is clear that the innovation process paradigm recognizes S&T as more of a reaction to societal needs and in many instances even as a function of economic issues, as shown in the competitiveness debate during the first half of the 1990s (Porter 1990, D'Andrea Tyson 1992, Krugman 1996). In Hungary, the first steps towards the acknowledgment of this policy paradigm have found their presence in the publications of the OMFB (1998).

See table 8.1 for a schematic comparison of the four time periods, the leading models of technological change and the predominant underlying notions of S&T policies during these times. This table is an enhanced version of table 1.1, which was presented in the introduction. Information on models of technological change and policy paradigms has been added. A number of factors seem to indicate that the models of technological change and S&T policy paradigms in fact are international phenomena. They seem to dominate S&T policy-making during similar time periods in all advanced industrial societies (compare with Blume 1985, Hall 1993, Ruivo 1994).

Table 8.1   S&T Paradigms, Policies and International Historic Environment, 1870s–1990s

Time	Internationally Leading Model of Technical Change	Internationally Predominant S&T Policy Paradigm	International Historic Setting
1870s–1940s	linear model	science push: S&T as source of national prowess and military strength	breaking up of the old order of Europe, coming to power first of capitalism and later of fascism
1950s–1960s	linear model	science push: S&T as the motor of progress and of military strength	bipolarism, Cold War
1970s–1980s	linear model	demand pull: S&T as a problem solver targeting societal and economic problems	intermittent weakening of bipolarism, Vietnam and Afghanistan Wars, oil and debt crises
1990s–	complex holistic model (including feed-backs and loops)	innovation process: S&T as a source of strategic opportunity, particularly for national and regional economies (both in the short and long term)	(economic) trilateralism, fall of the iron curtain, economic confrontation gradually replaces military confrontation

Source: A number of papers and books on 'waves', 'phases' and 'paradigms' of S&T policies, which have been influencing this categorization are listed in the section 'definition' of the annex.

In addition to what has been said, a few general lessons can be drawn with regards to the nature of policy paradigms. One of the research questions of this study targets the mechanisms of change applying to policy paradigms. Indeed, change seemed to be associated with sustained non-suitability of the dominant paradigm with the problems the policies are supposed to tackle. During the first transition, from the War economy to the Soviet S&T system in the 1940s, the policy paradigm was adapted and subject to incremental changes, i.e. policy instruments were partially changed and the goals of the S&T system were rearranged – for example the technical and some natural sciences were fostered more than before. During the second transition, to the NEM in the 1960s, the old policy paradigm was dropped as the S&T system for an extended time-period was not capable of producing the kind of results expected from it. Similarly, the policy paradigm changed again during the mid-1990s, as the Hungarian S&T system increasingly was perceived as not being competitive enough for the new world order. Therefore the change in policy paradigms was less a direct result of the political changes – these brought only less funding, but not a decisive modification of funding mechanisms – but, more a result of the full reopening of the Hungarian S&T system towards Western Europe, where many countries had changed their respective policy paradigms only shortly before.

In this, as in the other cases of policy paradigm change, the source of change was endogenous – it came from abroad. But in each case, decision-makers' trust in the virtues of the existing policy paradigm seems to have diminished considerably before the change actually occurred. Therefore, the main motivation for the changes was the perceived inability of the S&T system to aid the catch-up process with Western Europe and the recognition of the discussion of new policy paradigms on the international level.

Moreover, there seem to be large differences between the mechanisms of change within different policy fields. Policy fields, which are more contested than S&T policy, in many cases develop also policy paradigms – as was the case with Keynesianism and Monetarism in macroeconomic policy-making – but these paradigms often are subject to other influences, such as partisan differences. This is less likely to be the case in policy-fields, which are perceived to be technical and less partisan (compare Braun 1997, p.376) and which are therefore less likely to be subject to what sometimes colloquially is called 'high-politics' (Peterson/Sharp 1998, pp.163). Several authors have pointed out that these policy-fields frequently are quite complex and produce high levels of uncertainty about policy outcomes (Jacobsen 1995, Braun 1997). However, the validity of this argument is questionable, since typical cases of high-politics are similarly complex and also produce uncertainties – a good example are macro-economic policies.

In the 130 years subject to analysis in this study, the Hungarian attempts to economically close in on Western Europe were only partially successful. With the latest transition to capitalism the country attempted to reach a higher level of economic development. Regardless of the path Hungarian society chooses for the future, it seems clear that a reform of the S&T system would be beneficial for the prosperity of the national economy. For further steps in their efforts to restructure the S&T system, scholars, policy makers or any other interested group of persons have the opportunity to learn from similar periods of transition of the past. Consequently, before taking a look at possible policy options, the history of Hungarian S&T policies shall be evaluated here: an effort, which was indeed the third major research question posed at the beginning of this book.

The period, which demonstrates most forcefully the major lessons to be drawn from the study, is the years of transition from the fascist war economy to Stalinist central planning. The transfer of the Soviet S&T systems to Hungary in the early 1950s was less than effective. The realsocialist government of the time considered compartmentalization of research, education and production to be the key to rapid industrial development. This compartmentalization led to a number of problems: it is the cause of an educational system that was either out of touch with reality, as was the case in the discipline of economics, or over-specialized,[2] as was the case in technical disciplines. Furthermore, the compartmentalization led to the development of R&D institutions, which could not react to actual problems, either because the problems were unknown to the researchers or the researchers' solutions were unknown to or not accepted by industry. This, amongst other factors, led to the production of goods, which were of inferior quality in comparison to products from market economies.

In addition, the free flow of information was hamstrung by the authorities in centrally planned economies. In the long run, this free flow of information is a necessary precondition for a thriving economy and the development of S&T.[3] In the centrally planned realsocialist countries there was an inherent necessity to suppress the free flow of information in order to exert political control over society. The control of information was of major importance for the organization of society and the economy, both of which were dominated by hierarchical organizational structures. By integrating society vertically, the flow of information was also mostly vertical. Due to a number of reasons a horizontal information exchange began to develop during the 1980s. Partially responsible was the development of new information technologies as copiers, computers and fax machines. When the centrally planned realsocialist regimes were not capable of controlling the flow of information anymore, their authority began to crumble (Skolnikoff 1993, pp.96).

The flow of information was not only hamstrung inside the realsocialist countries, but also between the Council for Mutual Economic Assistance (COMECON) countries and the rest of the world. The political goal of an autarchic economy is therefore another factor in the explanation for the inadequacy of the policies of the realsocialist economies. Self-sufficiency should have lead to an eradication of the need to engage in trade with capitalist societies thereby circumventing possible economic and technological dependencies. Under these circumstances technology transfer was for the first decades of the People's Republic of Hungary possible only within the COMECON countries, which in many sectors had not reached high stages of technological development. Even later, when Hungary considered itself an open economy (Kádár 1984, pp.236) and engaged itself into overt and covert forms of technological transfer, imports from capitalist countries had to be kept at a moderate level because of the lack of hard currency reserves.

An important factor explaining the steering problems the realsocialist Hungarian governments had with regards to S&T, but more generally also the economy, is the disconnection of S&T on the one and economic policies on the other hand. This is partially due to the simplistic conception of innovation as a linear process, which found its expression in the science-push and demand-pull policy paradigms and encouraged policy-makers and administrators to isolate policy fields from each other.[4] Moreover, this understanding of the innovation process fits well with the similarly linear understanding of economic, social, scientific and technological development as put forward by Marxist-Leninist thought (Marx 1858), but also modernization theory (Apter 1965, Rostow 1962), both of which follow the path pursued by England at the beginning of the industrial revolution. Both view this form of development as the (sole) solution to the problem of relative economic backwardness.

Examples for the missing connection between S&T and economic policies can be found during all time periods analyzed in the course of the study. Most striking is, once again, the transformation of Hungary following the communist take-over in the late 1940s. The aggressive development of heavy industries did not affect the expansion of Hungary's already existing technological capacities, nor was it paired with a sensible accompanying S&T strategy. Although the decision-makers recognized that S&T policy was involved in the effort to further heavy industries, both policy fields were not integrated.

An example is the rising emphasis on the technical sciences in the Academy of Sciences as well as the universities. However, this academic emphasis was not followed by an effort to actually transfer the knowledge from the S&T system into industry. In addition, between 1949 and 1957 about 40 new industrial R&D institutes were established in Hungary, constituting the creation of a whole new S&T subsystem, which was highly valued by the political elite. At the same time company in-house R&D was to remain neglected. Consequently, the researchers responsible for R&D in the firms were less qualified and the capacity of the respective R&D units was meager. This is one of the explanations for the low rate of knowledge transfer from the S&T institutions to the companies.

The reforms of the 1960s, which were mostly part of the economic reform initiative New Economic Mechanism (NEM), were not able to remove the limitations on the Hungarian S&T system, which were caused by the decoupling of S&T and economic systems and policies. The effects of the NEM were described earlier. In short, the economic sphere was introduced to the notion of profitability resulting in the creation of a market controlled by a bureaucracy. The S&T system was confronted with policies that should have forced its institutions to act as if they too were in a market. Moreover, the budgetary constraints of the S&T system were greater than the ones the economic sphere was facing, since the NEM financing mechanisms forced researchers to seek collaboration with industry as a basic condition for their economic survival. However, because of the behavior of the industrial companies, the S&T institutions never were able to perform as under market conditions. As it turned out, the managers of industrial companies were more interested in pleasing the authorities and sticking to the rules of the bureaucracy than in behaving according to the rules of the market. They frequently refrained from implementing innovations, which in the short term, could have disrupted the management of their companies (Inzelt/Havas 1993).

Another important factor explaining the problems of S&T, but also of the economy at large, was that during the institutional transformation of the S&T system in the late 1940s and early 1950s policy makers destroyed the structures of the system which existed before the foundation of the realsocialist People's Republic. Linkages between the production sector and universities were destroyed, personnel was reshuffled from one part of the S&T system to the other, a number of the leading individuals in Hungarian science were forced either to leave the field or the country. Most scientific contacts existing before the late 1940s were cut because they mostly consisted of links to the capitalist world. The specific development of Hungary was not taken into account by the new regime. In hindsight, there is no other developmental strategy discernible for the first years of the People's Republic of Hungary than a blind following of the path the Soviet Union had taken under Stalin.

Therefore an important explanation for the failure of Hungarian S&T system reform efforts, but also partially of the economic policies of the 1950s, was that the decision-makers disregarded the differences between the socio-economic framework in which the set of policies they were about to employ in Hungary originally were constructed and the systemic specifications of Hungary, which often were very different from the former country. Again, the concept of the social system of production is helpful in explaining the processes, which led to the failure of the imported S&T policies. Note, the notion of social systems of production places the institutions of a country, whether they are firms, ministries, chambers of commerce or social clubs, in a context of ordering principles, for example the market, the state, the association or the network. These principles are prevalent in certain societies and are contingent on and embedded in a set of rules, norms and values, again highly specific to a certain society at a certain point in time.

As might be inferred from this brief description, the concept itself is highly contingent on the variables of time and space and emphasizes strongly the individual construction of each single social system of production. Accepting the presuppositions of this model, it is rather unlikely that a new measure or institution will fit smoothly into the new context, if it is transferred from one country to another. It is much more likely that the policy or institution is going to be misunderstood by the policy-makers who are deciding upon the policies or the administrators who actually carry out the policies in the country, where it has been transferred. A more benevolent possibility is that the institution or measure is changed by the new system (Boyer 1997), often to an extent, which changes its actual functions or operating procedures completely. S&T related examples range from the institutional transfer of the Humboldt-inspired German university model to the United States in the 19th century, the transfer of the German Kaiser-Wilhelm Institute to the Soviet Union in the early 20th century, and the transfer of the idea of Japanese precompetitive R&D consortia to the European Union, which resulted in the creation of the Framework Programmes for Research, Technological Development and Demonstration in the late 20th century.

Finally, one can sum up the result of the inquiry into the third major research question of the study, namely the divergence of the actual outcomes of the Hungarian S&T policies from the original intentions of decision-makers. During the time of realsocialism, from which most of the examples discussed in the last pages were drawn, this divergence is explained by a number of factors, amongst which are the following: the isolation of the different sectors of the economy, the compartmentalization of S&T, higher education and production systems, the disconnection of S&T and economic policies, the suppression of the free flow of information, the autarky strategy and the decision-makers' disregard of the S&T system's history. Moreover, these factors led to a low rate of innovation throughout the S&T system, the economic system and society at large. For the centrally planned economies a low rate of innovation meant that they only could adapt to changes in their environment very slowly. Ultimately, this may have been among the most important reasons for the demise of the People's Republic of Hungary.

Yet, it is important to understand that these problems are not limited to the time span from the 1940s to the 1980s, but that similar problems can be discerned through all the analyzed time periods. Clearly, in the periods before WWII the economic and science policies were not synchronized, and the same holds true for most Hungarian policies in the 1990s. Seeing that these problems are common to all industrialized countries the following section will apply the experiences of the Hungarian decision-makers to current policy-making.

8.2 What Can be Learned for Actual Policy-Making?

As clear as the shortcomings of the policies during the various time periods and the explanations for these inefficiencies may be, the question after the lessons to be drawn for actual policy-making remains to be answered. A few suggestions shall be presented here. Essentially there are five propositions, which address communication structures, policy coordination and integration, institutional and policy transfer, education, foreign direct investment and industrial R&D. Whilst some suggestions are directed at Hungary or at Central and East European Countries (CEECs) in transition from realsocialism to capitalism, a number of remarks are of a more general nature.

Moreover, most proposals are not directed at particular policy fields, such as technology or industrial policy, but are directed at specific problem sets. It is assumed that due to the increasing complexity of problems societies currently face, the standard categorization of issues according to pre-existing governance structures is not appropriate any more. These structures were set up to deal with another kind of problems stemming from a less internationalized world. Therefore, although after a more conventional denomination the discussed policies would fall under the categories of science, technology, educational and industrial policies, in absence of a better term most of them shall be addressed as 'knowledge policies'. This seems to be justified because the policies discussed here, all in one or another form address the knowledge base of advanced industrialized countries.

Since the late 1980s the rapid reconfiguration of scientific, technological and industrial governance structures in the OECD member countries may be interpreted as a sign for such a rethinking on the side of the respective governments. This reconfiguration of governance institutions can also be understood as an effect of the rise of the latest S&T policy paradigm and the model of technological change it is based upon. This insofar as a new policy paradigm signals a new way of thinking about a specific problem set and therefore requires a new set of institutions, which are to carry out the new policies.

Communication

The historical analysis of the Hungarian S&T system provides ample material for underscoring the importance of functioning communication structures for public policies. A case, which elucidates the problems that arise from inefficient communication between societal subsystems, is the S&T system during the time of realsocialism and here especially under Stalinism. As has been said before, one of the major problems of the S&T system, after the reorganization in 1949, was the compartmentalization of the system. Indeed, one might say that the radical idea of the rationalization of the production process, which stood behind the compartmentalization, failed miserably.

The functionality of communication processes has acquired more and more importance as an issue of policy-making during the last decade. One indication is the change from the linear to the holistic model of technological change, the latter of which attributes much more importance to communication processes than did the former. Another indication is the impact the book 'The New Production of Knowledge' by Michael Gibbons and others (1994) had on discussions about S&T policies. Yet, despite all the attention the book raised, it might still be underrated, as it offers lessons for policy-fields, which transcend the borders of S&T or even knowledge policies.

In this book the authors describe ongoing changes in the form of knowledge generation. These are, amongst others, signified by the increased production of knowledge in the context of application, a growing transdisciplinarity of knowledge generation and the proliferation of heterogeneous groups of researchers, called hybrid fora, where people from different disciplines exchange ideas.

The increasing application context of knowledge formation destroys the artificial boundaries between basic and applied science. Not only does this change the way S&T are perceived, but also the ways research is financed and institutions governing S&T are structured. Moreover, both financing as well as governing of research shall be influenced by the proliferation of heterogeneous groups of researchers, manifesting themselves in the constant birth, extinction and rebirth of groups with problem-centered missions, flatter hierarchies, denser networking qualities and transdisciplinary composition and visions. These heterogeneous groups are signified by the large percentage of people educated in different disciplines (and having learned to perceive the world in a specific way), who later came into contact with other forms of thinking and now are 'fluent' in several disciplines. It is interesting to notice that many firms already had to undergo similar changes (Biegelbauer 1997) when national economies internationalized in the 1970s and new forms of flexible specialization (Piore/Sabel 1984) arose in the 1980s.

All these characteristics of the new production of knowledge depend heavily on the abilities of a S&T system to create networks – this being the case for firms and research institutions alike. As the argument of Gibbons et al goes, the denser the networks are woven, the more efficient the knowledge production and with it the national system of innovation will be.

A number of other research projects have re-emphasized the importance of networking normatively (European Commission 1995), but also empirically (Hicks 1996, Müller et al 1996). Moreover, research on the factors of success in science over the last century have pointed out the importance of linkages between scientists and scientific institutions with diverse backgrounds, such as different forms of institutions and research work as well as different research fields (Hollingsworth 1998, Hage 1998). Furthermore, research of innovation systems has produced indications for the importance for successful research of research institutes' linkages to institutions from different societal (sub)systems (Müller et al 1996, Biegelbauer 1997).

If indeed these research results are correct, a major goal for knowledge policies should be the facilitation of linkages and networks, which themselves become actors and flexible storage units of knowledge. In fact, such policies can be momentarily discerned in a number of countries (European Commission 1997, OECD 1998) and international initiatives as the EU Framework Programmes. Unfortunately, in most Central and East European Countries (CEECs) these kind of policies are not common.

Due to the legacy of the hierarchical organization of state and society and in comparison to West European and North American advanced industrialized countries still relatively weak civil societies, the networks existing in CEECs are rather widely knit and primarily of a personal nature (Grabher 1993). This means that communication is sub-optimal insofar as the thinly knit personal networks are not institutionally based and therefore rather unstable. Moreover, a certain danger is involved in the prevalence of personal and weakness of institutional networks: the opaqueness of what might be called 'old boy networks' sometimes can be barely distinguished from corruption.

Coordination and Integration

The debates about the institutional set-up of European and North American countries' S&T systems take a lot of room in public discussions – unfortunately the form seems to be more important than the content in most of these cases. The time period most strikingly representing this problem in Hungary is the time of the latest transition to capitalism. The National Committee for Technological Development (OMFB), which in the Hungarian People's Republic was an independent bureaucratic unit, was formally raised to the level of a technology ministry in the first post-socialist governments of premier ministers Antall and Boross. However, in 1994, the next government decided to give the institution a semi-autonomous status, subjecting its most important policy instrument, the Central Technological Program (KMÜFA), to the Ministry of Industry, Trade and Tourism's (IKM) general economic fund. Similarly the first post-socialist government, after the beginning of the latest transition, created a science ministry, that then was abolished by the Horn government. These developments found their parallels in the functions of the Academy of Sciences, which at one time was responsible for administering the National Scientific Research Program (OTKA) and during another was not and then again was charged with the task, and many other functions of the Hungarian S&T system.[5] Similar changes were carried out in Austria and Germany, where the science ministries have changed their main functions (and their names) during the 1990s three times each, as well as in many other OECD countries.[6]

It seems to be a question of culture and even personal taste how a certain policy field is represented at the highest level of governance and if one believes a certain policy area to fit better into the agenda of a certain bureaucratic unit or another. In fact, this becomes clear in the discussion of the arrangement of ministries for Science, Education, Higher Education, Research, Technology, Trade, International Trade, Industry, Health, etc. which every few years, in many countries, becomes the center of political disputes (Golden 1991). The cycles of these discussions seem to coincide with the national elections of the very countries plagued by these discussions.[7] Indeed, the problem of institutional arrangements has been solved almost arbitrarily in different countries. Regrettably, there is no obvious best practice discernible, only a large number of individual solutions.

The same can be said about the coordination of national S&T efforts. For the fulfillment of coordination functions in advanced industrial economies intergovernmental committees are the 'best practice' solution. The Science Policy Council of the Antall/Boross governments, as well as the Horn administrations' Science Policy Collegium seem not to have been able to fulfill the task of efficiently coordinating S&T policies. It may however come as a small relief for the Hungarians that they share this particular problem with many other industrialized nations such as the Austrian Council for Science and Research, the FRG's Science Council and the United States' Federal Commission for the Coordination of Science and Technology, all of whom have a more or less effective advisory function, but are not institutions where actual coordination of policies takes place.

The coordination of economic and S&T policies is also relevant in this context and was outlined in the previous section. It has been shown that in the case of Hungarian S&T policies a major problem was the disconnection of the S&T and the economic systems. This was only partially addressed by policy-makers. In fact, even in these cases when policy-makers addressed the rift between the two systems, more often than not efforts to integrate economic and S&T policies were weak or not carried out. This non-integration might be caused by the intricate interest representation of the S&T and economic systems on the level of government, causing different patrons to defend their clients' interests so intensively that they lose sight of the interests of the whole S&T system or national economy. This ineffective integration might also be caused by unintended side-effects of policies resulting from the complex set-ups of S&T systems. An example, which is discussed in chapter 5, is the Hungarian economic reform program NEM that provided research institutes with a number of incentives consistent with the respective demand-pull policy paradigm. The reason why the measures were unsuccessful is to be found in the reaction of industry, which was not provided with corresponding incentives to engage in cooperation with research institutions. This was a clear case of insufficient integration or coordination of S&T and economic policies.

Accepting that the more applied side of academia and the rather R&D oriented sectors of industry can profit from each other, it seems to make sense to have them working together. In this case public policies can have a place in facilitating the cooperation: in fact, a number of countries have set their goals accordingly. What sometimes seems to hinder sensible policy formulation is the often artificial boundaries set between different forms of basic and applied research as well as industrial development, all of which are based on the still influential linear thinking of the science push or demand pull policy paradigms. These boundaries are often reflected in the split between science, technology and economic ministries, which tend to reify the very boundaries they want to overcome, if one is to believe the goals set in a number of S&T policy documents (for example BMFT 1993, BMWV 1996, DTI 1997, IKM 1997).

Institutional and Policy Transfer

Another lesson to be learned from the past transitions of the Hungarian S&T system is the importance of historically grown structures. As has been pointed out already in section 8.1, this is best exemplified by the first transition of the S&T system, during which the Soviet S&T structures were copied. The ignorance of realsocialist governments of the historically grown structures of the Hungarian national economy resulted in a disfigured and inefficient S&T system. For the latest transition back to capitalism it is essential to keep in mind that application-ready policy packages are dangerous as long as they simply transfer institutions and ignore the environment of the respective country in transition – this indeed is a dangerous aspect inherent to all attempts to identify and imitate 'best practice' models in any policy field.

Accepting the suppositions of the 'social systems of production' notion outlined before one must take into account where a 'best practice' model stems from and where it is about to be introduced. The differences between value structures and norm systems of the two countries must be considered. Then these institutions which are likely to interact frequently with the imported institution or which are the target of the imported policy will have to be looked at. It is entirely possible that the imported institution or policy will have a different effect on the social system of production into which it is transferred and that by way of feedback cycles, it will be necessary to adapt the institution or policy or its immediate environment.

An example, which links together the two subsections on communication as well as coordination and integration, is the communication between applied science and basic science as well as between science and development and engineering dominated institutions. The issue at the heart of the discussion is the linkage of S&T and economic policies. The standard best practice model for this type of communication facilitation is the introduction of an intermediary institution, such as the

'Fraunhofer Gesellschaft' in the FRG. This umbrella organization of institutes is financed by industry and government and aids the technology transfer between academia and industry. As has been pointed out earlier, the Hungarian Zoltán Bay Institutes were created after the German model (Pungor 1994, p.13). Only a few years after the foundation of the Zoltán Bay Institutes it is still too early to say if the Fraunhofer Institutes were a well-chosen blueprint for Hungary. Yet it seems obvious that the efforts of the institutes to integrate themselves into the Hungarian S&T system, by means of establishing close contacts to universities and firms alike, (see chapter 6) are increasing their likelihood of success.

Obviously, there exists a wide variety of additional institutional models for communication facilitation and knowledge transfer, ranging from university led development areas such as the Massachusetts Institute of Technology's (MIT) Route 128, to governmental extension services performed by the United States' National Institute of Standards and Technology (NIST), and information services as the Austrian Economic Chamber's foreign trade offices.

For a small open economy such as Hungary, which heavily depends on foreign trade, the latter organization seems to be specifically interesting. The Austrian foreign trade offices or the similar Japanese External Trade Organization (JETRO) have been useful instruments for obtaining information about issues as copyright practices, technical norms, import regulations and the like.[8] With the creation of the S&T attachés, Hungary has implemented a step into this direction (Nyiri 1994). Similarly, efforts to bridge the gap between academia and industry have led to the science-park of the technical university in Budapest and a number of other plans concerning similar institutions. Since the European Commission is further increasing its role as an information broker of S&T policy, it is to be expected that Hungary's EU accession will lead to a number of S&T policy innovations for the country.

Education

One of the keys to economic prosperity is the education of the population, as the outstanding strides of the East Asian Newly Industrialized Countries (NICs) – and the economic slump of other NICs less prone to foster education, as for example Argentine, Brazil or Syria – show.[9] In this respect it is interesting to notice that the CEECs have an excellent base to build upon. In the case of Hungary, the education system has reached international standards already early in the 20th century. This can be seen by the dozens of highly successful Hungarian scientists, who emigrated to Germany and to the United States during the 1920s and 1930s to help in the building of the atomic fission bomb (Leo Szilárd), the atomic fusion bomb (Edward Teller) or the computer (John Neumann). One even was awarded a Nobel prize in chemistry (Albert Szent-Györgyi). The quality of basic and higher education alike certainly played a role in the creation of so many extraordinary scientific careers.

Another more recent indication of the quality of Hungarian education is the successful integration of Hungarian foreign-led industry into international production networks (Zysman / Schwartz 1998, Czaban / Henderson 1998). Hungary was quite fast in capturing foreign direct investment (FDI), leading to the production of goods in Hungary, such as the Audi factory in Györ. This was possible because the German company, which is a division of Volkswagen, quickly understood that the work force in Györ had the skills necessary for building engines and other high value-added products. At least two factors can be identified that might have contributed to this successful integration of Hungarian industry into an international production network. One factor is the quality of the educational system. Another factor is the history of Hungarian automobile and truck production, which was quite successful in the Council for Mutual Economic Assistance (COMECON), but also in markets of developing countries, before it was rendered obsolete by the changes caused by the latest transition to capitalism.

Since the beginning of the latest transition to capitalism, the good education of the CEE work-force has produced a somewhat paradoxical situation. Because of the high level of education of their scientists and engineers some of the best and brightest (and youngest) have left the CEECs and become part of the brain drain. Yet, without a qualified body of researchers and engineers the future of the Hungarian national economy looks bleak. A rise in project-oriented funding of S&T might help to counter the brain drain, which at the moment, is affecting the most mobile layers of the researcher community.

Giorgi et al (1998) have shown that, while the brain drain of CEEC scientists going to foreign countries exists, the brain waste of scientists leaving the S&T system is much larger and potentially more destructive to the S&T system. It is important to notice that in many cases CEEC societies are not profiting from brainwaste – hence the name of the phenomenon – as most scientists and engineers leaving the S&T system are not taking positions in which their qualifications are used productively. Therefore, brain waste is not a cure against the shortage of university graduates and especially Ph.D. holders in an economy.

In its early stages is another way to work against brain drain and brain waste: the provision of funding for postgraduate research. It seems likely, however, that the current extent of the programs is not sufficient.[10] Funding of postgraduate work is also a way to foster the applicability of scholarly work to industry, as is exhibited by the Zoltán Bay network. Setting priorities in certain disciplines and the subsequent increase in funding in these fields recognized as of special importance will lure people into these areas.

For the future of the S&T systems of the CEECs it is important to notice that younger generations seem to feel the impact of realsocialism much less than do generations raised and educated in the times of the People's Republics. This is significant insofar as it is precisely the younger generations, who are grossly under-represented in the S&T systems of the CEECs. If the exclusion of younger researchers is not corrected in due time, a complete generation is going to be missing in the institutional set-up of the respective S&T systems. The ramifications of such a fact will be evident in the innovativeness of institutions, which will be missing the influx of new ideas embodied in young people.

Finally, it is a widely accepted truism that education does not stop with the entrance of a graduate into business life. The buzz-word addressing this notion, life-long learning, has to be achieved through trainee programs for the private as well as the public sector. National know-how from universities, the Academy of Sciences (MTA) or the Association for Innovation as well as international know-how through EU-programs might be utilized.

FDI and Industrial R&D

During the last years, a widely debated issue of knowledge policies concerned foreign direct investment (FDI). In a nutshell, the question is: Should FDI be welcome or not? On the one hand, the highly successful industrial countries of Japan, Taiwan and South Korea kept the levels of foreign capital in their countries low.[11] These countries found CEEC followers in the Czech Republic and Slovenia, who are also skeptical of FDI. On the other hand, it is clear that FDI often brings the winds of international competition, which are beneficial for a national economy. After all, multinational companies (MNCs) have accumulated tremendous know-how: they account for about four fifths of the privately financed R&D undertaken by OECD countries.[12] All over the world, the discussion over FDI divides countries and their strategies.

A concrete fear of those advocating restrictions against FDI is that too much foreign capital could deskill a national system of innovation by swamping the economy with transplants and other forms of low-skill production forms. A counter-argument would be that taking into account the discussion of FDI above, it is highly likely that opening up to the international environment might well be unavoidable in a time of increasing globalization pressures. This, of course, is all the more true for a country such as Hungary, which depends to a large extent on exports. Small advanced industrial countries should find it difficult to prosper economically without being woven into international production networks. Of course, the successful economic development of national economies is possible without the help of foreign capital, as has been shown by Japan, Taiwan and South Korea. However, it might well be that the East Asian NICs used the Cold War to generate profits and develop their economies by taking the side of NATO and receiving aid from the United States. This window of opportunity was closed in 1989 and therefore is not available for the CEECs anymore.

The decision to open Hungary to foreign capital, which was taken already in the early stages of the latest transition, has some tradition. As has been shown in chapter 3, the first wave of Hungarian industrialization in the 1870s was financed primarily with Austrian, but also German, capital. Over the following decades it was possible for Hungary to build an indigenous capital base and indeed it has been shown that in the 1910s industry was much less dependent on FDI than during the decades before.

Extending the direct basis of the question, one not only has to think about FDI proper, but in more general terms about the legacy of industry and industrial R&D institutions in the CEECs. These look back on a troubled past and face a still much more shaky future. In all countries, one problem that urgently needs to be addressed in some way is the industrial R&D system. The solution of Hungarian policy-makers in the 1990s has been to invite foreign capital investment into industry and, if possible, also into the industrial R&D institutes, which formerly was the centerpiece of the Hungarian industrial R&D system.

Accepting that the realsocialist industrial R&D institutes were too large and not modern enough for the new needs of the country, it is still open for debate exactly how much industrial R&D the country needs. After all, the industrial R&D capacities of Hungary are not so large that the country could single-handedly dismiss the industrial R&D institutes. They have been evaluated already. Based on these evaluations it should be possible to decide which institutes should, for strategic reasons, be kept under state control and which should be liquidated or privatized. The worst decision would be the implementation of a non-policy, i.e. to dry the industrial R&D institutes out by simply ignoring them, without making privatization possible.

With the premise that the existence of the industrial R&D institutes is desirable, there are two feasible options: one that government pumps money into the institutes and keeps them as transfer institutions or second that government decides to sell the institutes off. The second option, the sell-off, is the one which all administrations since 1990 have been insisting on implementing – without however doing so. While from the viewpoint of the government the wish to retain a stake in the R&D institutes is understandable, potential buyers from industry might have a different point of view. Therefore, it might make sense to sell the whole institute, and not to keep 26% or 51% of the shares. The first option, to pump money into the institutes and keep them, might be unrealistic because of general budgetary constraints. Moreover, it is even questionable if the R&D institutes would be capable of delivering what industry really needs. After all, the institutes would have to adapt to the market situation, in which they would have to approach firms and not the other way around. Under these circumstances survival in a stand alone option might be difficult for the institutes. It might be sensible therefore, to find an institutional affiliation for those institutes, which are not (yet) sold off.

Moreover, a reduction of potential patrons for the industrial R&D institutes could be helpful (compare with section 6.4). The State Assets Management Company (ÁV Rt.) certainly never was a good choice for any company to stay with for an extended period of time, especially for a R&D institute with the need to invest into its future. The institutes may be supervised by the National Committee for Technological Development (OMFB), perhaps in the form of new participants in the Zoltán Bay network

(Mosoni-Fried 1995, p.796), or by the Ministry for Trade and Industry (IKM).

Finally, on a more general level, pertaining to a number of the remarks made above: The national expenditures on R&D in all CEECs are not sufficient to retain the core competencies of the CEE S&T systems (EU 1997). In Hungary, it was clear from the outset that the goal of OMFB to reach an expenditure of 1.5% of the GDP for 1995 (Pungor 1994, p.15) was highly optimistic. However, the distance between actual expenditures and the goal persistently grew over the 1990s. The Orban government seems to be willing to foster the R&D expenditures of industry, but as long as general investment of industry into intangibles is as low as it is now, the expectations for support of R&D by industry should not be too high. Consequently, government should raise its support for the S&T system through available policy instruments such as the competitive project-based tenders of OMFB, until industry is back on its feet again. The alternative is to accept losses to the technology base of the country, which will negatively affect the future growth of the national economy.

Ultimately, the success or failure of efforts by Hungarian and other CEEC policy makers to build economic and political structures and achieve growth and living standards comparable to the economically most highly developed countries will not depend entirely on the success of institutional borrowing. Neither will it be a direct effect of radical reforms with the end of installing capitalism or socialism in any archetypal form. It will be a direct function of the ability of government to facilitate an interaction between a variety of societal interests, reacting to each others needs and ideas constructively. Such a form of organization is virtually unthinkable without an S&T system producing ideas for the solution of problems of society, in its economic, political and social subsystems. The success of the latest transition of Hungary is as dependent on the construction of a productive S&T system as the other transitions in the history of the country have been. If the policy makers are ready to learn from the past this transition might be the most successful in the history of modern Hungary.

Notes

1   Compare with the system theoretical notion of symbolically generalized exchange media and the norms and values on which the social systems of production are ultimately contingent on, both described in section 2.2.

2   Loren Graham tells the story of his first visit to the Soviet Union in 1960, when he met a young woman, who he asked about her profession. As it turned out, she had earned a degree as an engineer in the field of ball-bearings for paper mills. See Graham 1993, p.30.

3   On the role of the information flow for the development of S&T and the economic growth of West Europe, see Rosenberg/Birdzell 1986.

4   The innovation management literature shows that similar pitfalls are known to company managers, too (Iansiti, Marco and West, Jonathan, 1997).

5   Compare with chapter 6.

6   For Austria, see section 7.3.

7   This hypothesis has been proven, again, in the months before the elections to Parliament in Austria in the fall of 1994 and 1995 as well as in early 1997, when chancellor Klima took office. All of a sudden discussions about new configurations of the Science and Research, Economic Affairs, Transport, Education and Culture resorts broke out – only to vanish again, shortly after their materialization, when other issues began to dominate public discussion.

8   It is interesting to notice in this respect that the US limited itself to the present day to a similar structure as Hungary does, namely attachés in a handful of embassies of key importance for the US. However, since the end of the Cold War public again and again discussions in the US about the future of the Central Intelligence Agency (CIA) have led to arguments about the possibility to use the CIA as an organization engaging more strongly into data gathering and the provision of services for the economy.

9   See, among others, for accounts about the role of education in the successful efforts of South Korea, Taiwan, Singapore and Hong Kong, Wade 1990, Vogel 1991, Chi-Ming/San 1993 and Kim 1993.

10   Several persons I talked to shared this opinion. Some of them were students, i.e. they had a direct interest in expanding funds for postgraduate education, whereas others had no linkages to the higher education system.

11   South Korea, however, changed its policies with regards to foreign direct investment in the early 1990s with the result that the country was much more directly effected by the Asian financial crises of mid-1998 than either Taiwan or Japan.

12   Dunning, J.H., 'Governments, Economic Organisation and International Competitiveness', Discussion Paper, vol II (1989/90), Reading, University of Reading, cited from: Jaklic 1994.

Annex

Realsocialism

The term 'realsocialism' is used here as a description for the real-world experience of the Council for Mutual Economic Assistance (COMECON) countries which were using Marxism-Leninism as their state-ideology, but were only partially structured according to the ideas of Karl Marx. Still, the Socialist People's Republics were named after Karl Marx's and Vladimir Iljitsch Lenin's theories of 'Socialism,' i.e. the era that should lead or, according to Marx's own conviction, should have led already during the 19th century – via the dictatorship of the masses – to 'Communism', the era of classless society. In the strict sense of original Marxist theory Communism never existed on earth. Therefore, the usage of the term 'Communism' is not an apt description for the People's Republics, which were perhaps 'socialist', or to signify the aberration of these systems from Marx's own thoughts, 'state socialist' or 'realsocialist'.

The state ideology of the COMECON countries, Marxism-Leninism, has been modified in a number of ways from the original thought of the founders of Marxism, Karl Marx and Friedrich Engels, and even the first General Secretary of the Communist Party of the Soviet Union, V.I. Lenin. The primary goal of the modifications was to make the ideology fit the needs of the Communist Party. A classic example is the above mentioned expectation of Marx that the revolution of the proletariat would happen in the near future, i.e. the second half of the 19th century. Soon after the death of Marx, when it became clear that this was not to happen, Marxists explained this with a variety of reasons. In the first decade of the 20th century the Austromarxist Rudolf Hilferding classified financial capital as the highest form of capitalism. This notion was furthered by Lenin, who in 1917 introduced the theory of 'imperialism', the most elaborate form of 'monopoly capitalism'. Lenin insisted that imperialism, as it could be observed in the 1910s, was the last stage of capitalism leading finally to its decay (Lenin 1950).

As it turned out after WWII, capitalism was even more adaptable than Hilferding, Lenin and others had thought. The communist parties in a number of states in the West were crushed by the stronger bourgeois parties believing in the superiority of market based economic organization. To explain this, a new term was introduced, namely 'late capitalism'. According to the ideological line of the Soviet Union and some Western European communist parties, late capitalism was characterized by the final betrayal of the working class by the Social Democrats. Capitalism was seen as capable of disciplining the proletariat through the Keynesian welfare state, which was constructed mainly by the Social Democrats and of which the Social Democratic unions were an integral part (Abendroth 1965, 1995; Fülberth 1994; Fritzsche 1996).

There have been numerous other examples for such add-ons to Marxist theories generated to explain the development of capitalism not foreseen by Marx, Engels and, later, Lenin. The term 'realsocialism' has been used regularly in Western Europe as a description of the system of the Eastern Bloc countries, but scarcely in the United States and never in the Eastern Bloc until 1989.

Science

Science is knowledge about the universe that is obtained in systematic and reproducible ways according to a specific set of methods (usually, but not necessarily, seen as congruent with or closely related to the 'scientific method' as developed by Francis Bacon and Isaac Newton). Moreover, science is also the process of gathering the information, which then is classified, interpreted and related to each other so as to be transformed into knowledge; this process again is subject to rigorous regulation.

Technology

Technology is knowledge about ways and means to fulfill human purposes in a specifiable and reproducible manner (compare with Brooks 1980). This knowledge is not bound to tangible goods: the knowledge of the production, as well as the use of the goods are part of the term technology. This includes forms of organization as well as processes of production. It is important to notice that the terms science and technology are not congruent, although both deal with knowledge and are frequently used together (compare Skolnikoff 1993, pp. 12).

S&T System

In contrast to the 'innovation system', (Lundvall 1985, 1992, Nelson 1993, Niosi 1993, Niosi/Bellon 1994, Wijnberg 1994) a somewhat more all-encompassing term, the 'Science & Technology' (S&T) 'system' used here, consists of the research, development and engineering (R, D & E) as well as the educational institutions of a national economy, regardless of their private, public, profit or non-profit nature, and the political structures directly related to these organizations. 'Directly related' refers to political superstructure such as councils, ministries and agencies that supervise the research, development and engineering (R, D & E) and educational institutions. Excluded from the term S&T system are organizations such as unions, chambers of commerce or associations in which the institutions or the institutions' members are members, which would be included in the broader term 'innovation system'. An exception to this rule are intermediary institutions, whose primary function is aiding the knowledge transfer between basic and applied science as well as industry and academic institutions, which fulfill important functions in most S&T systems. For an elaboration on the 'systems' notion, see, earlier in the study, the section on systems theory and S&T systems (2.2).

Industrial R&D system

The term 'industrial research and development' (R&D) system here is used for defining an institutional structure which has the purpose to plan and engage in industrial R&D. Parallel to the definition of the S&T system, besides the R&D institutions themselves, engineering-related organizations and the applicable political superstructure are also included in this term. 'Directly related', again, refers to institutions such as councils, ministries or agencies related to the above mentioned institutions from the industrial sector. The industrial R&D system is part of the larger S&T system.

Paradigm

The term has acquired new meaning in the social sciences through Thomas Kuhn. Kuhn writes in his book 'The Structure of Scientific Revolutions' (1970, p.10):

> In this essay, 'normal science' means research firmly based upon one or more past scientific achievements [the bases for a paradigm, or perhaps 'Weltanschauung', P.B.] achievements that some particular scientific community acknowledges for a time as supplying the foundation for its further practice.

He provides evidence by citing Aristotle's world view, later overcome by Newton, who created a world view which in itself was overcome by Einstein. Two characteristics are shared by these achievements, he expounds (1970, p.10).

> Their achievement was sufficiently unprecedented to attract an enduring group of adherents away from competing modes of scientific activity. Simultaneously, it was sufficiently open-ended to leave all sorts of problems for the redefined group of practitioners to resolve. Achievements that share these two characteristics I shall henceforth refer to as 'paradigms', a term that relates closely to 'normal science'.

Kuhn's term here is applied to traditions in S&T policies that are, if less monumental than his examples, still paradigmatic in the sense of distinct cognitive frameworks, which form the basis of a new tradition in S&T policy, including new overarching goals, new instruments and new instrument settings (Hall 1993). It is important to notice that a set of S&T policies here are not understood by themselves as a paradigm – different from Beatriz Ruizo's (1994) understanding of the notion. Here a policy paradigm is a network of underlying assumptions about a set of S&T policies, which are then shared and used by an increasing number of decision-makers in a specific policy-field. Policy paradigms provide knowledge on how to structure the world, interpret phenomena related to a specific problem-set and policy-field. Thereby, policy paradigms pre-select possibilities of political action, primarily with regards to the ends, but partially also the means to reach these ends. Policy paradigms are specific, small-scale views of the world, which do not address the whole world as does a Weltanschauung, rather a limited part of it such as the innovation process in the case of S&T policies or the nature and function of the market in economic policies. Thereby policy paradigms are also important for the actual policy-making process, as they create inter-subjective meaning for persons and groups.

This understanding of the term policy paradigm is quite similar to Peter Hall's (1986, 1989, 1993) definition of the concept. Yet, as Hall's notion was created to account for the impact of economic theories on economic policy-making, with France and Great Britain as case studies illustrating the changes from Keynesian to Monetarist macroeconomic policies, a few differences in scope and nature of the terms seem to apply. First, Hall's notion of policy paradigms as 'ideologies', defined as 'well developed networks of ideas, which prescribe a course of economic action' (1986, p.278) may overextend the usage of the term. Indeed, one of the goals of the concept of policy paradigms is to make the concept independent of classical ideologies such as Weltanschauungen (Hall 1986, p. 278).

A helpful aspect of Hall's (1993) work is his elaborate distinction between the three different orders of change. First order changes signify changes in policy setting, second order changes are changes of policy instruments and setting and third order changes apply to changes of the former two and the overarching goals of policies. For this study only third-order changes are of interest with respect to a policy paradigm's effects. These, however, are understood as consisting of two levels, the first of which pertain to changes in policy paradigms, such as the science-push to the demand-pull paradigm (see below). The second level pertains to changes in the notions on which the paradigms themselves are based, i.e. the models of technological change. These form the basis, but are not congruent with, the policy paradigms. Extending the terminology of Peter Hall, these might be termed fourth order changes.

Another facet that is important, is that actual S&T policies might be partially outside the currently dominant policy paradigm. This is also a clear difference to Kuhn's notion of a paradigm that did not entail the parallel existence of paradigms for prolonged periods of time. In the case of policy paradigms in the S&T policy field, the 'demand pull' paradigm never was able to eliminate block funding as a significant source of income for research institutions in any economically developed country, including Hungary.

The names of the first and second of the paradigms used here, 'science push' and 'demand pull' have been used by Stuart Blume (Blume 1985, p.2), who distinguished the different periods of Dutch science policy with these chapter headings. Beatriz Ruivo used the terms nine years later when she suggested what is known as the 'phases' of science policies, which might better be described as 'paradigms' of science, or here S&T, policies (Ruivo 1994). 'Science push' and 'demand pull' refer to the 'pipeline model', in which innovation is seen as consisting of a chain of events from basic research, to applied research, to development, to engineering and finally to production. In the 'science push' paradigm basic research is the leading force responsible for progress, in the 'demand pull' paradigm it is the demand of the market, which pulls the research with it. The third, more complex paradigm, tries to circumvent this oversimplification. This paradigm is still developing, but lacks at the moment a singular unifying notion comparable to the role the 'pipeline model' had for the two preceding paradigms. Here, it is referred to as the 'innovation process' paradigm, which itself is based on the complex holistic model, whose name is a reference to the more complex nature of the model, overcoming linear thinking and including feed-back loops (Kline/Rosenberg 1986).

Glossary

CEE	Central and Eastern Europe
CEECs	Central and Eastern European Countries
CEFTA	Central European Free Trade Association
COCOM	Coordinating Committee for Export to Communist Area
COMECON	Council for Mutual Economic Assistance, also: CMEA
EU	European Union, until 1992 European Communities, EC
FEFA	(Felsöoktatásfejlesztési Alap)
	Higher Education Fund
FRG	Federal Republic of Germany
GDP	Gross Domestic Product
GFA	(Gazdaságfejlesztési Alap)
	Economic Development Fund
HUF	Hungarian Forint
ICSU	International Council of Scientific Unions
IKM	(Ipari És Kereskedelmi Minisztérium)
	Ministry for Industry and Trade, later Tourism was added
KMÜFA	(Központi Müszaki Fejlesztési Alap)
	Centralized Technological Development Fund
MNC	Multinational Corporation, also: Transnational Corporation, TNC
MSZMP	(Magyar Szocialista Munkáspart)
	Hungarian Socialist Workers Party, Communist Party of Hungary
MTA	(Magyar Tudományos Akadémia)
	Hungarian Academy of Sciences
MÜFA	(Müszaki Fejlesztési Alap)
	[Company] Technical Development Fund
NEM	New Economic Mechanism
NPO	Non-Profit Organization
OECD	Organisation for Economic Cooperation and Development

OKKFT	(Országos Kozéptavú Kutatási Fejlesztési Terv) National Medium Term R&D Plan
OMFB	(Országos Müszaki Fejlesztési Bizottság) National Committee for Technical Development, later National Committee for Technological Development
OTKA	(Országos Tudományos Kutatási Alap) National Scientific Research Fund, later National Scientific Research Program
OTTKT	National Long Range Plan for Scientific Research
R, D&E	Research, Development and Engineering
S&T	Science and Technology
STS	Science, Technology and Society Studies
SMEs	Small and Middle Enterprises
TBP	(Tudománypolitikai Bizottság) Science Policy Committee
USD	US Dollar
USSR	Union of Socialist Soviet Republics
WWI	World War I
WWII	World War II

Bibliography

Abendroth, W. (1965), *Sozialgeschichte der Arbeiterbewegung* (Social History of the Workers' Movement), Frankfurt.

Abendroth, W. (1995), *Einführung in die Geschichte der Arbeiterbewegung* (Introduction Into the History of the Workers' Movement), Heilbronn.

Abott, A. (1994), 'Hungarian Coalition has Pro-Science Leanings', *Nature*, September 1.

Aichholzer, G. Martinsen, R. and Melchior, J. (1994), *Österreichische Technologiepolitik auf dem Prüfstand* (An Analysis of the Austrian Technology Policy), Research Paper 13, Political Science Department, Institute for Advanced Studies, Vienna.

Albert, M. (1993), *Capitalism vs. Capitalism*, Four Wall Eight Windows.

Almond, G. and Verba, S. (1963), *The Civic Culture*, New Jersey, Princeton University Press.

Apter, D. (1965), *The Politics of Modernization*, The University of Chicago Press.

Arrow, K. (1971, 1.ed 1962), 'Economic Welfare and the Allocation of Resources for Invention', in Rosenberg, Nathan (ed), *The Economics of Technological Change – Selected Readings*, Penguin Books Ltd, pp. 179.

Bachmaier, P. (ed) (1991), *Bildungspolitik in Osteuropa* (Education Policy in Eastern Europe), Jugend & Volk.

Balassa, B. (1982), *The Hungarian Economic Reform*, World Bank Staff Working Paper No. 506, 2/82.

Balázs, K. et al (1990), 'The Management of Research and Development in Hungary at the End of the 80s', *Soviet Studies*, 4/90, pp. 723–741.

Balázs, K. (1992), 'Transition of the Science and Technology Management System in Hungary', *Science and Public Policy*, 2/92, pp. 89–97.

Balázs, K. (1993), 'Lessons from an Economy with Limited Market Functions: R&D in Hungary in the 1980s', *Research Policy*, vol. 22, pp. 537–552.

224

Balázs, K. (1994), *Transition Crisis in Hungary's R&D sector*, Institute of Economics of the MTA, Budapest, Typescript.

Balázs, K. and Plonsky, G.A. (1994), 'Academy-Industry Relations in Middle-Income Countries: East Europe and Ibero-America', *Science and Public Policy*. April.

Balázs, K. (1995), 'Innovation Potential Embodied in Research Organizations in Central and Eastern Europe', *Social Studies of Science*, vol. 25, 4/95, pp. 655.3.1

Bay Zoltán Foundation (1995), *Bay Zoltán Foundation for Applied Research*, Budapest.

Bay Zoltán Foundation (1998), *Bay Zoltán Foundation for Applied Research*, Budapest.

Becker, G. (1976), *An Economic Approach to Human Behaviour*, Chicago, Chicago University Press.

Bell, D. (1973), *The Postindustrial Society*, New York.

Bentley, A. (1949, 1.ed. 1909), *The Process of Government*, Bloomington, Ind., Principia Press, in L. D'Andrea Tyson (1992), *Who is Bashing Whom? Trade Conflict in High Technology Industries*, Institute for International Economics, Washington.

Berend, I. and Ránki, G. (1960), *The Development of the Manufacturing Industry in Hungary*, Studia Historica, 19/60, MTA.

Berend, I. and Ránki, G. (1985), *The Hungarian Economy in the Twentieth Century*, Croom Helm.

Bessenyei, I., Debreczeni, P. and Setenyi, J. (1994), 'Report on Hungarian Education', in *Issues in Transition, Transformation of the National Higher Education and Research Systems of Central Europe*, vol. 7., IWM, Vienna.

Bessenyei, I. and Melchior, J., (1996), *Die Hochschulpolitik in Österreich und Ungarn 1945–1995*, Peter Lang.

Biegelbauer, P. (1990), *Foreign Participation in Austria's Economy*, prepared for the international project: Europe in the Time of the World Economic Crises, University of Vienna, Typescript.

Biegelbauer, P. (1994), *Evaluation of the Effects of the OECD Report on 'Science, Technology and Innovation Policies in Hungary' on the Country*, Report for the OECD Technology Audit for Hungary, MIT.

Biegelbauer, P. (1996), 'Realizing Hungary's Potential – The Country's Industrial R&D System in Transition', *Eastern Europe Working Paper Series*, Institute for Advanced Studies (IAS), Vienna.

Biegelbauer, P. (1996), 'The Innovation System of Slovenia', *EU-PHARE Program Study,* Working Paper Series of the Interdisciplinary Center for Comparative Research in the Social Sciences (ICCR), Vienna.

Biegelbauer, P. and Filzmaier, P. (1996), 'Geheimrezepte: Die US-Technologiepolitik unter Clinton' (Secret Formulas: US-Technology Policy under Clinton), *Salzburger Nachrichten,* November 16th, Science Part, p. V.

Biegelbauer, P. (1997), 'To Be Innovative or Not to Be – Innovative Behavior of Austrian Companies', *Wirtschaftspolitische Blätter,* Austria, 5/97, October, pp. 517.

Biegelbauer, P., Giorgi, L. and Pohoryles, R.(1998), 'Research and Technological Development Cooperation Activities of the EU/EEA Countries with the Central and Eastern European and Baltic States in the Field of Scientific and Technological Research', *Working Paper Series of the ICCR,* Vienna, Austria.

Biegelbauer, P. (1998), 'Forschungs- und Technologiepolitik in Slowenien und Ungarn' (Science and Technology Policy in Slovenia and Hungary), in *Wirtschaftspolitische Blätter,* 4/98, pp. 399.

Bijker, W. (1995), *Of Bicycles, Bakelites and Bulbs: Toward a Theory of Sociotechnical Change,* MIT Press.

Blume, S. (1985), *The Development of Dutch Science Policy in International Perspective, 1965–1980,* RAWB, pp.2.

Blyth, M. M. (1997), 'Any more Bright Ideas? The Ideational Turn of Comparative Political Economy', in *Comparative Politics,* vol. 29, 2/97, pp. 229–250.

BMFT (1993), *Bundesbericht Forschung,* Bonn.

BMWV (1993), *Technologiepolitisches Konzept der österreichischen Bundesregierung,* Austrian Ministry for Science and Transport, Vienna.

BMWV (1996), *Technologiepolitisches Konzept der Bundesregierung,* (Technology Policy Concept of the Federal Government), Vienna.

BMWV (1999), *Bericht des Bundesministers für Wissenschaft und Verkehr an den Nationalrat: Österreichische Forschungsstrategie, Phase I* (Report of the Federal Ministry for Science and Transport to the National Assembly: The Austrian Research Strategy, Phase I.), Austrian Ministry for Science and Transport, Vienna.

Bouché, P. (1998), 'Alternative Approaches to Industria R&D Institutes in Hungary and Russia', in Meske et al (1998), *Transforming Science and Technology Systems – The Endless Transition?,* IOS Press, pp. 183.

Boskin, M. and Lau, L. (1992), 'Capital, Technology and Economic Growth', in N. Rosenberg et al, *Technology and the Wealth of Nations*, Stanford University Press, pp. 32.

Boyer, R. and Hollingsworth, R. (eds) (1997), *Contemporary Capitalism: The Embeddedness of Institutions*, Cambridge and New York, Cambridge University Press.

Brabant, van, J. (1993), *Industrial Policy in Eastern Europe*, Kluwer Academic Publishers.

Braun, D. (1997), *Die politische Steuerung der Wissenschaft*, Frankfurt am Main, Campus.

Braun, T. et al (1994), 'World Science in the Eighties', *Scientometrics*, Vol. 29, 3/94, pp. 299.

Braun, T., Schubert, A. (1996), *Indicators of Research Output in the Sciences From 5 Central European Countries, 1990–1994*, paper prepared for the UNESCO meeting Basic Sciences for Development: Subregional Opportunities and Challenges (Central Europe), held in January, Keszthely, Hungary.

Brooks, H. (1980), 'Technology, Evolution and Purpose', *Daedalus 109*, 1/80, pp.65–81.

Business Central Europe (1995), 'Focus: Hungary', December, pp. 28.

Budapest Business Journal (1996), 'Top 30 Corporate Foreign Investors', November 25 – December 1, pp. 8.

Business Central Europe – The Annual (1996/97), 'Hungary'. Business Central Europe – The Annual.

Campbell, D. (1993), *Strukturen und Modelle der Forschungsfinanzierung in Deutschland – Eine Policy-Analyse*, (Structures and Models of the Financing of Research in Germany – A Policy Analyses), Forschungsbericht, Institute for Advanced Studies (IAS), Vienna.

Campbell, D. and Felderer, B. (1994), *Forschungsfinanzierung in Europa*, Manz.

Cardoso, F.H. (1971), *Ideologias de la burguesia industrial en sociedades dependientes (Argentina y Brasil)*, Mexico.

Central Statistical Office (1995), *Hungarian Statistical Yearbook 1994*, Budapest.

Central Statistical Office (1996), *Foreign Direct Investment in Hungary 1994*, Budapest.

Central Statistical Office (1996), *Statistical Handbook of Hungary 1995*, CSO.

Chesler, C. (1994), 'Infighting slows Push to Privatization', *The Budapest Sun*, Aug. 25–31, pp. 1–2.

Chiang, J.T. (1990), 'Management of Technology in Centrally Planned Economies', *Technology in Society*, 12/90, pp. 397–426.

Chilosi, A. (1992), 'Market Socialism: A Historical View and a Retrospective Assessment', *Economic Systems*, 1/92, pp. 171.

Chi-Ming, H. and San, G. (1993), 'National Systems Supporting Technical Advance in Industry: The Case of Taiwan', in R. Nelson (ed), *National Innovation Systems*, Oxford University Press.

Clark, C. and Lam, D. (1994) *Taiwan's Experience with Privatization and its Implications for the Former Soviet Bloc*, Pacific Basin Research Center, Center for Science and International Affairs, Harvard University, Typescript.

Cohen, B. (1985), *Revolution in Science*, Cambridge, Mass., Belknap Press.

Csöndes M. and Räty, T. (1985), 'Financing of Research and Development', in K.O. Donner, and L. Pál (eds), *Science and Technology Policies in Finland and Hungary – A Comparative Study*, Akadémiai Kiadó.

Czaban, L. and Henderson, J. (1998), *Globalization, Institutional Legacies and Industrial Transformation in Eastern Europe*, Paper prepared for the Annual Conference of the Society for the Advancement of Socio-Economics held in Vienna in July, Austria.

Darvas, G. (ed) (1988), *Science and Technology in Eastern Europe*, Longman.

Darvas, G. (ed) (1995), 'Transformation of the Science and Technological Development System in Hungary', in R. Mayntz (ed), *Transformation Mittel- und Osteuropäischer Wissenschaftssysteme – Länderberichte*, Leske + Budrich, pp. 853–977.

David, P. and Foray, D. (1994), 'Accessing and Expanding the Science and Technology Knowledge Base', *DSTI/STP/TIP*, 4/94, OECD.

Donner, K. and Pál, L. (eds) (1985), *Science and Technology Policies in Finland and Hungary – A Comparative Study*, Akadémiai Kiadó.

Downs, A. (1957), *An Economic Theory of Democracy*, New York, Harper & Row.

DTI/ OST (1995), *White Paper on Competitiveness – Forging Ahead*, London.

Dunning, J. H. (1994), *Governments, Economic Organisation and International Competitiveness*, Discussion Paper, vol. II (1989/90), University of Reading, cited in: Jaklic, M. (1994), *Concepts for an Industrial Policy: The Case of Slovenia*, WIIW-Working Paper,Vienna.

Easton, D. (1965), *A Framework for Political Analysis*, Englewood Cliffs, New York.

EBRD (1997), *Transition Report 1997,* Great Britain.

Eisenstadt, S.N. and Rene L. (eds) (1981), *Political Clientelism, Patronage and Development*, London, Sage.

England, J.M. (1982), *A Patron for Pure Science – The National Science Foundation's Formative Years, 1945–57*, NSF.

Etzkowitz, H. (1996), *Academic-Industry-Government Relations: Corporatism American Style*, Paper presented at the Universities and the Global Knowledge Economy: A Triple Helix of University-Industry-Government Relations Conference in Amsterdam, 3–6 January.

European Commission (1994), *The European Report on Science and Technology Indicators 1994*, EU.

European Commission (1995), *Greenbook on Innovation*, EU.

European Commission (1997), *Yearly Report on R&D Activities for 1996*, Kom (97) 373. July.

Farkas, J. (1985), 'The State of the Art in Science and Technology Policy Studies', in K.O. Donner and L. Pál (eds), *Science and Technology Policies in Finland and Hungary – A Comparative Study*, Akadémiai Kiadó.

Fetscher, I. (1965), *Der Marxismus: Seine Geschichte in Dokumenten*, (Marxist History in Documents), München, Piper-Verlag.

Field III, F. and Clark, J. (1997), 'A Practical Road to Lightweight Cars', *Technology Review*, January, pp. 28.

Fischer, H. and Szabadváry (1995), *Technologietransfer und Wissensaustausch zwischen Ungarn und Deutschland*, R. Oldenbourg Verlag.

Foerster, H. (1984), 'Principles of Self-Organization', in H. Ulrich, G. Probst, *Self-Organization and Management of Social Systems*, Berlin.

Frankel, M. and Cave, J. (1997), *Evaluating Science and Scientists*, Central European University Press.

Freeman, C. (1982), *The Economics of Industrial Innovation*, 2nd ed, MIT Press.

Freeman, C. (1987), *Technology Policy and Economic Performance*, London, Pinter.

Fritzsche, K. (1996), *Sozialismus*, in: Neumann, F., *Handbuch Politische Theorien und Ideologien 2* (Handbook of Political Theories and Ideologies 2), Opladen.

Fülberth, G. (1994), *Der große Versuch* (The Big Experiment), Köln.

Füzeséri, A. (1992), *Über die Ungarn, in der Geschichte der Naturwissenschaft und Technik*, OMFB, Budapest.

Galeano, E. (1973), *Die offenen Adern Lateinamerikas*, Wuppertal.

Geipel, G. (1991), 'The Failure and Future of Information Technology Policies in Eastern Europe', *Technology in Society*, 13/91, pp. 207–228.

Gerschenkron, A. (1962), 'Economic Backwardness in Historical Perspective', in idem, *Economic Backwardness in Historical Perspective*, Harvard University Press.

Gerschenkron, A. (1977), *An Economic Spurt that Failed – 4 Lectures in Austrian History*, Princeton University Press.

Giorgi, L. (1997), *Brain Drain: The Emigration of Scientists from Relevant Parts of the NIS*, ICCR Project Report, INTAS Project 93–684. Vienna.

Glagow, M. and Willke, H. (1987), *Dezentrale Gesellschaftssteuerung*, Centaurus Verlagsgesellschaft

Golden, W. (1991), *Worldwide Science and Technology Advice to the Highest Levels of Government*, Pergamon Press.

Goldstein, J. and Keohane, R. (1993), *Ideas and Foreign Policy*, Ithaca, NY, Cornell University Press.

Gorz, Á. (1985), *Paths to Paradise: On the Liberation From Work*, South End Press.

Grabher, G. (ed) (1993), *The Embedded Firm: On the Socioeconomics of Industrial Networks*, Routledge.

Graham, L. (1993), *Science in Russia and the Soviet Union – A Short History*, Cambridge University Press.

Graham, L. (1993), 'Palchinsky's Travels', *Technology Review*, No. 6/1993, pp. 23–31.

Grande, E. and Häusler, J. (1994), *Industrieforschung und Forschungspolitik*, Köln, Campus Verlag.

Grossman, G. and Helpman, E. (1991), *Innovation and Growth in the Global Economy*, MIT Press.

Hage, J. (1998), *Organizational Innovation and Organizational Change*, Typescript.

Halasz, G. (1989), *Higher Education in Hungary*, OKI, Budapest.

Halasz, G. and Lukács, P. (1990), *Education Policy for the 1990s*, Hungarian Institute for Education Research.

Hall, P. (1986), *Governing the Economy: The Politics of State Intervention in Britain and France*, Oxford University Press.

Hall, P. (ed) (1989), *The Political Power of Economic Ideas: Keynesianism Across Nations*, Princeton, Princeton University Press.

Hall, P. (1993), 'Policy Paradigms, Social Learning, and the State', *Comparative Politics*, April, pp. 275– 296.

Handberg, R. and Xinming, L. (1992), 'Science and Technology Policy in China: National Strategies for Innovation and Change', *Technology in Society*, pp. 271.

Hangos, K. (1997), 'The Limits of Peer Review: The Case of Hungary', in M. Frankel, J. Cave, *Evaluating Science and Scientists*, Central European University Press.

Hanson, P. and Pavitt, K. (1987), *The Comparative Economics of Research Development and Innovation in East and West: A Survey*, harwood academic publishers.

Haslinger, P. (1996), *Hundert Jahre Nachbarschaft – Die Beziehungen zwischen Österreich und Ungarn 1895–1994*, (A Hundred Years of Neighbourhood – the Relations between Austria and Hungary 1895– 1994), Peter Lang Verlag.

Heilbroner, R. (1967), 'Do Machines Make History?', *Technology and Culture*, vol. 8, pp. 335–345.

Heinrich, H.G. (1986), *Hungary – Politics, Economics and Society*, Lynne Rienner Publishers.

Herman, Á. and Stubnya, G. (1994), *Information System in Transition – The Hungarian Scene*, Budapest, Typescript.

Hicks, D. and Katz, J. (1996), 'Science Policy for a Highly Collaborative Science System', *Science and Public Policy*, vol 23, pp. 39–44.

Hofmann, J. (1993), *Implizite Theorien in der Politik – Interpretationsprobleme regionaler Technologiepolitik*, Opladen, Westdeutscher Verlag.

Hollingsworth, J. R., Schmitter, P. and Streek, W. (eds) (1993), *Governing Capitalist Economies: Performance and Control of Economic Sectors*, Oxford University Press.

Hollingsworth, J. R. and Streek, W. (1994), *Governance of Capitalist Economies*, Oxford University Press.

Hollingsworth, J. R. (1998), *Doing Institutional Analysis: Implications for the Study of Innovations*, Mimeo.

Hrebenar, R. and Scott, R. (1990, 1.ed 1982), *Interest Group Politics in America*, Englewood Cliffs, Prentice Hall.

Hughes, T. (1983), *Networks of Power*, Baltimore, Johns Hopkins University Press.

Huntington, S. (1991), *The Third Wave*, London, Norman.

IKM (1997), *Information Pertaining to Technical Development and Technical Cooperation*, Working Paper of the Technology Policies Department.

Imre, J. (1998), 'S&T in Hungary', in W. Meske et al, *Transforming Science and Technology Systems – The Endless Transition?*, IOS Press.

Inzelt, A. (1982), 'Economic Sensitivity in Technological Development in Hungary', *Acta Oeconomica*, vol. 28 1–2/82.

Inzelt, A. and Havas, A. (1993), *Major Innovation and Technology Policy Issues in Hungary*, IKU, Budapest.

Inzelt, A (ed) (1994), *Institutional Support for Technological Improvement – Back to Cooperation from Rigid Seperation*, draft version, World Bank project, Institutional Support for Technological Improvement, IKM, Budapest.

Inzelt, A. (1994), 'Privatization and Innovation in Hungary: First Experiences', *Economic Systems*, June.

Inzelt, A. (1995), *Review of Recent Developments in Science and Technology in Hungary*, study for the OECD Technology Audit of Hungary, February, Budapest, Typescript.

Inzelt, A. (1995), *For a Better Understanding of the Innovation Process in Hungary*, STEEP Discussion Paper No. 22, June, University of Sussex.

Jacobsen, J. K. (1995), 'Much Ado about Ideas. The Cognitive Factor in Economic Policy', in *World Politics*, 47/95, pp. 283–310.

Jaklic, M. (1994), *Concepts for an Industrial Policy: The Case of Slovenia*, WIIW, Vienna.

Janos, A. (1982), *The Politics of Backwardness in Hungary*, Princeton University Press.

Jánossy, F. (1969), 'Gazdaságunk mai ellentmondásainak eredete és felszámolásuk útja' (The Origins of the Present Economic Problems of Hungary and the Way to Solve Them), *Közgazdasági Szemle*, July-August, pp. 806–829.

Jasanoff, S. et al (1995), *Handbook of Science and Technology Studies*, Sage, Thousand Oaks.

Kádár, J. (1984), *Sozialismus und Demokratie in Ungarn*, Budapest, Corvina Kiadó.

Kálmán, A. (1996), 'Von Reformen geplagt', *Österreichische Hochschulzeitung*, October, pp. 28.

Katzenstein, P. (1985), *Small States in World Markets: Industrial Policy in Europe*, Cornell University Press.

Kautsky, K. (1925), *The Labour Revolution*, George Allen and Unwin, London.

Keck, O. (1993), 'The National System for Technical Innovation in Germany', in R. Nelson (ed), *National Innovation Systems*, Oxford University Press, pp. 122.

Keller, C. et al (ed) (1985), *Technology, Politics and Economics: Papers Presented to the Fourth Swiss-Hungarian Roundtable, 15–19 October 1984*, EOEW.

Kennedy, P. (1987), *The Rise and Fall of the Great Powers*, New York, Vintage.

Keren, M. (1992), 'The New Economic System, the New Economic Mechanism, and the Yugoslav LMF: Bureaucratic Limits to Reform', *Economic Systems*, 1/92, pp. 89.

Kim, L. (1993), 'National Systems of Industrial Innovation: Dynamics of Capability Building in Korea', in R. Nelson (ed), *National Innovation Systems*, Oxford University Press.

Kinzer, S. (1994), 'A Wall of Resentment Now Divides Germany', *NYT*, October 14th, pp. A1, 14.

Kline, S. and Rosenberg, N. (1986), 'An Overview of Innovation: The Positive Sum Strategiy', in R. Landau and N. Rosenberg (eds), *Harnessing Technology for Economic Growth*, Washington, DC., National Academy Press.

Kornai, J. (1980), *Economics of Shortage*, two volumes, North-Holland Publishing Co.

Kornai, J. (1986, 1.ed 1983), *Dilemmas and Contradictions*, MIT Press.

Kornai, J. (1990), *Vision and Reality, Market and State*, Harvester Wheatsheaf.

Kornai, J. (1993), *Transformational Recession – A General Phenomenon Examined Through the Example of Hungary's Development*, Collegium Budapest/Institute for Advanced Studies (IAS), Discussion Papers No.1.

Kosáry, D. (1992), 'Academy and Reform: The Athenaeum Experiment', in *Strategies for Support of Scientific Research*, MTA/NSF.

Kregel, J., Matzner, E. and Grabher, G. (1992), *The Market Shock – An Agenda for the Economic and Social Reconstruction of Central and Eastern Europe*, Austrian Academy of Sciences (Research Unit for Socio-Economics).

Krugman, P. (1996), *Pop Internationalism*, MIT Press.

Kuhn, T. (1970, 1.ed 1962), *The Structure of Scientific Revolutions*, Chicago, Chicago University Press.

Lakatos, I. and Musgrave, A. (1970), *Criticism and the Growth of Knowledge*, Cambridge, Cambridge University Press

Landes, D. (1969), *The Unbound Prometheus*, New York, Cambridge University Press.

Lenin, V.I. (1950, 1.ed. 1916), 'Imperialism – The Highest Stage of Capitalism', in *Selected Works*, Foreign Languages Publishing House, Moscow, vol. 1, Book 2.

Lenin, V. I. (1984, 1.ed. 1917), *Staat und Revolution* (State and Revolution), Dietz, Berlin.

Lenin, V. I. (1988, 1.ed. 1913), *Drei Quellen und drei Bestandteile des Marxismus* (Three Sources and Three Parts of Marxim), Berlin, Dietz-Verlag.

Luhmann, N. (1984), *Soziale Systeme*, Suhrkamp.

Luhmann, N. (1990), *Die Wissenschaft der Gesellschaft*, Suhrkamp.

Lundvall, B. A. (1985), *Product Innovation and User-Producer Interaction*, Aalborg University Press.

Lundvall, B. A. (1992), *National Systems of Innovation*, Pinter, London.

Mádi, C. (1990), 'Transfer of Technology – Hungary in the Eighties: Hungary's Trade in Intellectual Products', *Acta Oeconomica*, vol. 42, Akadémiai Kiadó, Budapest.

Mark and Cave, J. (1997), *Evaluating Science and Scientists*, Central European University Press.

Marres, N. (1998), 'Did NASA become the Post Office gone to Space'?, *EAAST Review*, 18/98, pp.7–11.

Marx, K. and Engels, F. (1848), *Manifest der Kommunistischen Partei*, (Communist Manifesto), Berlin, Dietz-Verlag.

Marx, K. (1978, 1.ed. 1858), 'Die Grundrisse', in R. Tucker, *The Marx-Engels Reader*, 2nd ed., W.W. Norton & Co., pp. 221.

Mayntz, R. (ed) (1995), *Transformation Mittel- und Osteuropäischer Wissenschaftssysteme – Länderberichte*, Leske + Budrich.

McCormack, G. (1990), 'Capitalism Triumphant? The Evidence From Japan', *Monthly Review*, May, pp.1.

McCurdy, H. (1993), *Inside NASA, High Technology and Organisational Change in the US Space Program*, Johns Hopkins University Press.

Merton, J. (1982), *A Patron for Pure Science – The National Science Foundation's Formative Years, 1945–57*, NSF.

Meske, W. (1992), 'The Restructuring of the East German Research System', *Science and Public Policy*, 5/92, pp. 298–312.

Meske, W. et al (1998), *Transforming Science and Technology Systems – The Endless Transition?*, IOS Press.

Meth-Cohn, D. (1995), 'Hopes and Fears', *Business Central Europe*, February, pp. 7–8.

Mihályi, P. (1993), 'Hungary: a unique approach to privatization – past, present and future', in I. Székely and D. Newbery, *Hungary: an Economy in Transition*, Cambridge University Press, pp.84.

Ministry for Trade and Industry (1993), *Industrial Policy for the 1990s*, Ministry for Trade and Industry.

Ministry of Foreign Affairs (1991), *Economic Action Programme of the Hungarian Government*, Fact Sheets on Hungary No.2.

Ministry of Foreign Affairs (1991), *Hungary – Democracy Reborn*, Ministry of Foreign Affairs.

Ministry of S&T (1994), *Science in Slovenia*, MZT, Ljubljana.

Mosoni, J. (1994), 'Industrial Research Institutes in the Transition Period in Hungary', first version, in Inzelt, A. (ed), *Institutional Support for Technological Improvement – Back to Cooperation from Rigid Seperation*, World Bank project: Institutional Support for Technological Improvement, IKM, Budapest.

Mosoni, J. (1998), 'Structural Changes in Industrial R&D in Hungary: Losers and Winners', in W. Meske et al (1998), *Transforming Science and Technology Systems – The Endless Transition?*, IOS Press.

Mosoni-Fried, J. (1995), 'Industrial Research in Hungary: A Victim of Structural Change', *Social Studies of Science*, vol. 25, 4/95, pp. 777.

MTA/NSF (1992), *Strategies for Support of Scientific Research*.

MTA (several editions), *Newsletter of the MTA*, Number 1 (July 1991), Number 2 (June 1992), Number 3 (November 1992), Number 4 (April 1993), Number 6 (September 1994), Number 7 (February 1995), Number 12 (1995), MTA.

Müller, K. (1996), *The Austrian Innovation System*, Project Reports I–VII, Institute for Advanced Studies, Vienna.

Müller, L. (1985), 'Planning and Coordination in Science and Technology', in K.O. Donner and L. Pál (eds), *Science and Technology Policies in Finland and Hungary – A Comparative Study*, Akadémiai Kiadó, pp. 225.

National Academy of Sciences (1994), *National Academy Press*, National Academy Press, Spring.

Nelson, R., Winter, S. and Schuette, H. (1976), 'Technical Change in an Evolutionary Model', *Quarterly Journal of Economics*, vol 90, pp. 90–118.

Nelson, R. and Winter, S. (1982), *An Evolutionary Theory of Economic Change*, Cambridge, Harvard University Press.

Nelson, R. and Rosenberg, N. (1993), 'Technical Innovation and National Systems', in R. Nelson (ed), *National Innovation Systems*, Oxford University Press.

Nelson, R. (ed) (1993), *National Innovation Systems*, Oxford University Press.

Newsletter of The MTA (1993), 'Evaluation of Institutes of the HAS by Experts of ICSU', 4/93, MTA, pp. 3.

Newsletter of the Hungarian Academy of Sciences (1994), No. 6, September.

Niosi, J. et al (1993), 'National Systems of Innovation: In Search of a Workable Concept', *Technology in Society*, 2/93, pp. 207.

Niosi, J. and Bellon, B. (1994), 'The Global Interdependence of National Innovation Systems: Evidence, Limits and Implications', *Technology in Society*, 2/94, pp. 173.

North, D. (1981), *Structure and Change in Economic History*, New York, Norton.

North, D. (1990), *Institutions, Institutional Change and Economic Performance*, Cambridge and New York, Cambridge University Press.

Nyiri, L. (1994), 'S&T Diplomacy', *The Hungarian Observer*, vol. 7, 2/94, pp. 36.

OECD (1988), *Reviews of the National S&T Systems – Austria*.

OECD (1993), *Science, Technology and Innovation Policies: Hungary*, Centre for Cooperation with European Economies in Transition.

OECD (1997), *Main Science and Technology Indicators, 2/97*.

OECD (1998), 'New Rationale and Approaches in Technology and Innovation policy', *STI review special issue* No. 22.

Office for Science and Technology of the United Kingdom (1993), *Realising Our Potential – A Strategy for Science, Engineering and Technology*, White Paper, London.

OMFB (1991): *The National Committee for Technological Development of the Republic of Hungary*, Budapest.

OMFB/IKM/Ministry of Finance (1993), *Innovation Policy of the Hungarian Government*, Budapest.

OMFB (1994), *OMFB – National Committee for Technological Development*, Budapest.

OMFB (1995), *OMFB – National Committee for Technological Development*, Budapest.

OMFB (1995), *The Government's Technical Development Concept*, Proposal to the Government, Septmeber, Budapest.

OMFB (1996), *OMFB – National Committee for Technological Development*, Budapest.

OMFB (1998), *Research and Technological Development in Hungary*, Budapest.

OSI (1997), *Ost-Dokumentation*, vol. 11, 3/97, Österreichisches Ost- und Südosteuropa Institut.

OSI (1997), *Ost-Dokumentation*, vol. 11, 4/97, Österreichisches Ost- und Südosteuropa Institut.

Pacey, A. (1990), *Technology in World Civilization*, Cambridge, Mass., MIT Press.

Palló, G. (1995), 'Deutsch-ungarische Beziehungen in den Naturwissenschaften im 20. Jahrhundert', in H. Fischer and Szabadváry, *Technologietransfer und Wissensaustausch zwischen Ungarn und Deutschland*, R. Oldenbourg Verlag, pp. 277.

Papp, B. (1995, 1996), 'Survey Hungary', *Business Central Europe*, Dec. 1995/Jan. 1996, pp.20.

Paulinyi, A. (1995), 'Industrieförderung und Techniktransfer aus dem Deutschen Reich', in H. Fischer and Szabadváry, *Technologietransfer und Wissensaustausch zwischen Ungarn und Deutschland*, R. Oldenbourg Verlag.

Pavitt, K. (1992), 'Internationalisation of Technological Innovation', *Science and Public Policy*, 2/92, pp. 119–123.

Peck, S. (1991), *Research, Technology Transfer, and Information Flow in Former Soviet Bloc Nations: The Case of Hungary*, Dissertation, University of Maryland College Park.

Pécsi, K. (ed) (1993), *Selected Science Indicators of the Hungarian Academy of Sciences*, Secretariat of Research Policy of the Hungarian Academy of Sciences, Budapest.

Perlez, J. (1994), 'Fast and Slow Lanes on the Capitalist Road', *NYT*, October 7th, p. A1, p. A12.

Perlez, J. (1994), 'Once-Promising Hungary Struggles with Economic Slump', *NYT*, October 26th, p. A14.

Péteri, G. (1993), 'Scientists versus Scholars: The Prelude to Communist Takeover in Hungarian Science, 1945–1947', *Minerva*, vol. 31, 3/93, pp. 290–325.

Péteri, G. (1994), *On the Legacy of State-Socialism in Academia*, Paper presented at the 4th Conference of the International Society for the Study of European Ideas, The European Legacy: Towards New Paradigms, August 22–27, Graz, Austria.

Piore, M. and Sabel, C. (1984) *The Second Industrial Divide*, New York, Basic Books.

Piskunov, D. and Saltykov, B. (1992), 'Transforming the Basic Structures and Operating Mechanisms in Soviet Science', *Science and Public Policy*, 4/92, pp.111.

Polanyi, K. (1956, 1.ed 1944), *The Great Transformation*, Beacon, Beacon Press.

Porter, M. (1990), *The Competitive Advantage of Nations*, New York, Free Press.

Pribersky, A. (1996), 'Die Logik der Blöcke fortdenken'?, *Der Standard*, June 16th, pp. 23.

Pungor, E. (1991), *The Scientific and Technological Development System in Hungary*, Hungary.

Pungor, E. and Nyiri, L. (1993), 'The Reconstruction of Science and Technology in Hungary', *Technology in Society*, 15/93, pp. 25–39.

Pungor, E. (1994), *Strategie der Ungarischen Technologischen Entwicklungspolitik,* (strategy of the Hungarian technological developmental policy), Vortrag vor der Deutsch-Ungarischen Industrie und Handelskammer (talk given at the German-Hungarian Chamber of Industry and Commerce), January 31th, Budapest, Typescript.

Pungor, E. (1994), *Strategie der Ungarischen Technologischen Entwicklungspolitik,* (Strategy of the Hungarian Technological Development Policy), Vortrag vor der Deutsch-Ungarischen Industrie- und Handelskammer, (Talk given at the German-Hungarian Chamber of Industry and Commerce), in January 31st, Budapest, Typescript.

Rapoport, A. (1992), *Allgemeine Systemtheorie,* Verlag Darmstädter Blätter.

Reed, J. (1995/ 1996), 'The Great Growth Race', *Central European Economic Review,* December, pp. 8.

Richter, S. (1992), *The Transition from Command to Market Economies in East-Central Europe,* The Vienna Institute for Comparative Economic Studies, Yearbook IV, Westview Press.

Roemer, J. (1993), 'Can There Be Socialism After Communism'?, *Market Socialism,* Oxford University Press, pp. 80.

Rokkan, S. (1987), *Centre-Periphery Structures in Europe,* Frankfurt am Main, Campus.

Rosenberg, N. (ed) (1971), *The Economics of Technological Change – Selected Readings,* Penguin Books Ltd.

Rosenberg, N. and Birdzell, L.E. (1986), *How the West Grew Rich: The Economic Transformation of the Industrial World,* Basic Books.

Rostow, W. (1960), *The Stages of Economic Growth,* Cambridge University Press.

Rothschild (1971), *The Organization and Management of Government Research and Development: A Framework for Government Resarch and Development,* London.

Ruivo, B. (1994), 'Phases or Paradigms of Science Policy'?, *Science and Public Policy,* 3/94, pp. 157.

Sachs, J. and Woo, W. T. (1993), *Structural Factors in the Economic Reforms of China, Eastern Europe and the Former Soviet Union,* December (revised version), Paper presented at the Economic Policy Panel meeting in Brussels, October 22–23, Belgium.

Salomon, M. (1996), 'Leistungskontrolle für Forschung und Lehre', *Der Standard,* July 7th, pp. 5.

Scherer, F.M. (1984), *Innovation and Growth – Schumpeterian Perspectives*, MIT Press.

Schertler, W. (1988), *Unternehmensorganisation*, R. Oldenbourg Verlag. 3rd ed.

Schimank, U. (1987), 'Evolution, Selbstreferenz und Steuerung komplexer Organisationssysteme', in M. Glagow and H. Willke, *Dezentrale Gesellschaftssteuerung*, Centaurus Verlagsgesellschaft.

Schimank, U. (1995), 'Transformation of Research Systems in Central and Eastern Europe: A Coincidence of Opportunities and Trouble', *Social Studies of Science*, vol. 25, 4/95, pp. 633.

Schödl, G. (1990), 'Wissenschaft und Wissenschaftspolitik in Ungarn' (Science and Science Policy in Hungary), in G. Schödl, *Formen und Grenzen des Nationalen* (Forms and Boundaries of the 'National'), Institut für Gesellschaft und Wissenschaft, Erlangen, pp. 243.

Schmitter, P. (1979), 'Still the Century of Corporatism', in P. Schmitter and G. Lehmbruch (eds), *Trends Towards Corporatist Intermediation*. Beverly Hills, Sage.

Schumpeter, J. (1971, 1.ed 1928), 'The Instability of Capitalism', in N. Rosenberg (ed), *The Economics of Technological Change – Selected Readings*, Penguin Books Ltd.

Schumpeter, J. (1975), *Capitalism, Socialism and Democracy*, Harper & Row.

Secretariat of the Hungarian Rector's Conference (1994), *The Hungarian Rector's Conference*, March 21st, Hungary.

Senghaas, D. (ed) (1972), *Imperialismus und strukturelle Gewalt: Analysen über abhängige Reproduktion*, Suhrkamp.

Senghaas, D. (ed) (1974), *Peripherer Kapitalismus: Analysen über Abhängigkeit und Unterentwicklung*, Suhrkamp.

Sharp, M. and Peterson, J. (1998), *Technology Policy in the European Union,* New York, St. Martin's Press.

Simai, M. (1985), 'Research, Development, Innovation and Technology Flows in Small Countries: The Experience of Hungary', in C. Keller et al (ed), *Technology, Politics and Economics: Papers Presented to the Fourth Swiss-Hungarian Roundtable, 15–19 October 1984*, EOEW, pp. 57.

Simon, D. and Goldman, M. (eds) (1989), *Science and Technology in Post-Mao China*, Harvard University Press.

Simonis, G. and Martinsen, R. (1995), *Paradigmenwechsel in der Technologiepolitik?* (Paradigm Change in Technology Policy?), Opladen, Leske + Budrich.

Skolnikoff, E. (1993), *The Elusive Transformation – Science, Technology, and the Evolution of International Politics*, Princeton University Press.

Smith, A. (1974), *The Wealth of Nations*, Pelican Classics.

Smith, K. (1994), 'Interactions in Knowledge Systems', *DSTI/STP/TIP*, 15/94, OECD.

Smith, M. R. and Marx, L. (1994), *Does Technology Drive History?* MIT Press, pp. 53–66.

Spaulding, R. (1991), 'German Trade Policy in Eastern Europe', in *International Organization*, Volume 45, 3/91, pp. 343.

Der Standard (1994), 'Osteuropa driftet auseinander: WIIW Wirtschaftsforscher erwarten Aufsschwung für Ungarn, Slowakei und Tschechien', July 8th, pp. 23.

Der Standard (1996), 'Jeder dritte Betrieb will Teil der Produktion in den Osten verlegen', February 15th, pp. 19.

Der Standard (1996), 'Mobile Berater erleichtern Strukturwandel', Februar 27th, pp. 23.

Der Standard (1997), 'Ungarn's Premier gelobt: Keine weitere Belastung', January 9th, pp. 15.

Der Standard (1997), 'Regierung plant Offensive für die Exportwirtschaft', January 14th, pp. 14.

Stankovsky, J. (1996), 'Bedeutung ausländischer Direktinvestitionen in Osteuropa', in *WIFO Monatsberichte*, 2/96, pp. 123.

Stark, D. (1995), 'The Hidden Character of East European Capitalism: Recombinant Ownership', in *Transition*, World Bank, vol.6, 11–12/95, pp. 15.

Stein, R. (1987), *Centre Periphery Studies in Europe*, Campus, Frankfurt am Main.

Steiner, J. (1996), 'Der Bund verteilt die Technologiemilliarde', in *Der Standard*, December 12th, pp. 20.

Stewart, I. (1948), *Organizing Scientific Research for War – The Administrative History of the Office for Scientific Research and Development*, Atlantic Monthly Press.

Strassel, K. (1996), 'Park Fever', in *Central and Eastern European Business Review*, 11/96, pp. 18.

Streek, W. and Schmitter, P. (eds) (1985), *Private Interest Government*, Sage.

Stucke, A. (1993), *Institutionalisierung der Forschungspolitik*, Frankfurt am Main, Campus.

Szabadváry, F. and Vámos, E. (1994), *Tausend Jahre Technik und Innovation in Ungarn* (Thousand Years of Technology and Innovation in Hungary), Technisches Nationalmuseum, Budapest.

Székely, D. (ed) (1988), *The Hungarian Academy of Sciences*, HAS.

Székely, I. and Newbery, D. (1993), *Hungary: an Economy in Transition*, Cambridge University Press.

Tamás, P. (1985), 'Historical Processes of the Institutionalization of Science and Technology', in K.O. Donner and L. Pal (eds), *Science and Technology Policies in Finland and Hungary*, Akadémiai Kiadó, pp. 30–31.

Tarkowski, J. (1981), 'Poland: Patrons and Clients in a Planned Economy', in S.N. Eisenstadt and R. Lemarchand (eds), *Political Clientelism, Patronage and Development*, London, Sage.

Tarnóczy, M. (1992), *Hungarian Science – at Home and Abroad*, Institute for Research Organization of the Hungarian Academy of Sciences, Transcript.

Tolnai, M., Quittner, J. and Darvas, G. (1985), 'The Organizational Framework of Science and Technology Policy', in Donner and Pál (eds), *Science and Technology Policies in Finland and Hungary – A Comparative Study*, Akadémiai Kiadó.

Transition (1996), 'Quotation of the Month', July-August 1996, IMF-Washington DC., pp. 9.

Truman, D. (1971, 1.ed 1956), *The Governmental Process*, New York, Knopf.

Tucker, R. (1978), *The Marx-Engels Reader*, 2nd ed, W.W. Norton & Co.

Urban, W. (1992), 'Economic Lessons from Two Newly Industrializing Countries in the Far East'? in S. Richter, *The Transition from Command to Market Economies in East-Central Europe*, The Vienna Institute for Comparative Economic Studies, Yearbook IV, Westview Press, pp. 251.

Valdés, B. (1999), *Economic Growth*, Great Britain, Edward Elgar Cheltenham.

Vámos, É. (1995), 'Deutsch-Ungarische Beziehungen auf dem Gebiet der Chemie', in H. Fischer and Szabadváry, *Technologietransfer und Wissensaustausch zwischen Ungarn und Deutschland*, R. Oldenbourg Verlag, pp. 217.

Vas-Zoltán, P. (1985), 'Methods of Evaluation of R&D Achievements', in K.O. Donner and L. Pál (eds), *Science and Technology Policies in Finland and Hungary – A Comparative Study*, Akadémiai Kiadó, pp. 239.

Vogel, D. (1978), 'Why Businessmen Distrust Their State: The Political Consciousness of American Corporative Executives', in *British Journal of Political Sciences*, 8/78, pp. 45–78.

Vogel, D. (1986), *National Styles of Regulation*, Cornell University Press.

Vogel, E. (1991), *The Four Little Dragons*, Harvard University Press.

The Wall Street Journal's Central European Economic Review (1994), 'Techno-Trophies', vol.2, pp. 18.

Wade, R. (1990), *Governing the Market: Economic Theory and the Role of Government in East Asian Industrialization*, Princeton University Press.

Wagschal, U. (1996), *Staatsverschuldung im internationalen Vergleich*, Leske + Budrich.

Wald, M. (1994), 'U.S. to Aid Big 3 in Cleaner-Car Research', in *NYT*, October 19th, pp.D1/D16

Wang, Y.F. (1993), *China's Science and Technology Policy: 1949–1989*, Avebury Ashgate Publishing.

Wedel, J. (1994), 'Lessons of Western Technical Aid to Central and Eastern Europe', in *Transition – The Newsletter about Reforming Economies*, vol. 5, July/August, pp. 14.

Weiss, C. J. (1993), 'Scientific and Technological Responses to Structural Adjustment: Human Resources and Research Issues in Hungary, Turkey, and Yugoslavia', in *Technology in Society*, vol. 15, pp. 281.

Weyer, J. (1987), 'Reden über Technik als Strategie sozialer Innovation' , in M. Glagow and Willke, H., *Dezentrale Gesellschaftssteuerung*, Centaurus Verlagsgesellschaft.

Whitney, C. (1994), 'East Europe's Hard Path to a New Day', in *NYT*, September 30th, p. A1, p. A10, p. A11.

Wijnberg, N. (1994), 'National Systems of Innovation: Selection Environments and Selection Processes', in *Technology in Society*, 3/94, pp. 313.

Williams, R. (1994), 'The Political and Feminist Dimensionsof Technological Determinism', in M. Smith and L. Marx, *Does Technology Drive History?*, Cambridge, Mass., MIT Press, pp. 217–236.

Willke, H. (1987), *Systemtheorie*, Gustav Fischer Verlag, 2nd ed.

Wörgötter, A. et al (1995), *The Pioneer of Reforms – Hungary*, Study Commissioned by Bank Austria AG, Institute for Advanced Studies (IAS), Vienna.

Yee, A. S. (1996), 'The causal effects of ideas on policies', *International Organization*, 50/96, pp 69–108.

Zambra, K. (1994), *Wechselwirkung zwischen der Performance Art und der Politik in den USA* (The Relationship between Performance Art and Politics in the USA), Mag. Phil. Thesis, University of Vienna.

Zysman, J. (1983), *Governments, Markets, and Growth*, Cornell University Press.

Zysman, J. and Schwartz, A. (1998), 'Reunifying Europe in an Emerging World Economy: Economic Heterogeneity, New Industrial Options, and Political Choices', *Journal of Common Market Studies*, 36/83, pp.405–429.

Zysman, J. and Schwartz, A. (eds) (1998), *The Industrial Foundation of a New Political Reality*, University of California, Berkeley.

Index